Ralf Engbert

Dynamical Models In Neurocognitive Psychology

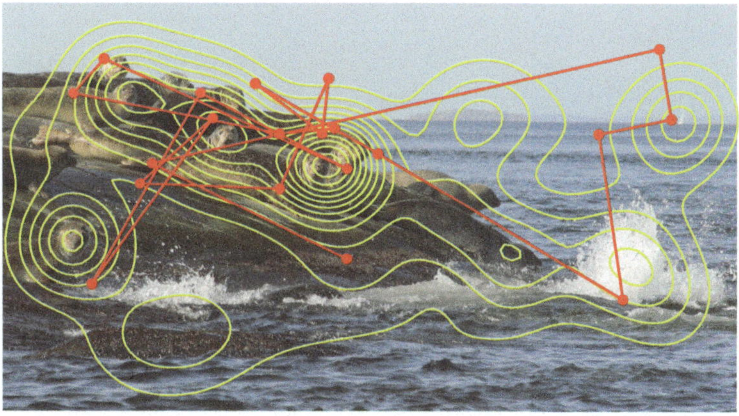

Springer

Ralf Engbert
Department of Psychology
University of Potsdam
Potsdam, Germany

ISBN 978-3-030-67298-0 ISBN 978-3-030-67299-7 (eBook)
https://doi.org/10.1007/978-3-030-67299-7

This Springer imprint is published by the registered company Springer Nature Switzerland AG
The registered company address is: Gewerbestrasse 11, 6330 Cham, Switzerland

To Hilde, Johannes, and Sebastian
—RE

Preface

The development of mathematical models is an important step in the challenging research program to improve our understanding of the complex relation between cognitive processes, activity in the nervous system, and human behavior. In this research, biological processes on the level of molecular and cellular interactions are as important as processes on the level of large-scale brain activity and the high-resolution observation of behavior using cutting-edge recording techniques—the instruments of cognitive science.

The aim of this textbook is to provide examples of the interplay between experimental research, data analysis, and mathematical modeling of human behavior—from the perspective of ongoing research in experimental and cognitive psychology. For the development of mathematical models, neurophysiological foundations play an increasingly important role, since our knowledge on biological implementations of processes of human information processing and control of action provides boundary conditions for model choice, so that biological plausibility is a very important criterion for model selection. Therefore, we will use the term *neurocognitive psychology* to indicate that mathematical models are often used for the integration of knowledge from neuroscience and cognitive science.

A common theme throughout the different chapters of the lecture is research on human motor control (mostly eye movements), attention, and visual perception. I expect that this focus on a limited research field helps to draw connections between the different mathematical topics.

The lecture starts with a short introduction to the generation of electrical activity in single neurons (Chap. 1). The microscopic coding of information via neural firing rates is mathematically described as a *stochastic point process* that has important applications to macroscopic processes such as saccade generation in experimental tasks, which is discussed in several chapters of the book.

The analysis of miniature eye movements and the underlying statistical properties is discussed in Chap. 2. If we fixate an object of interest, our eyes produce miniature eye movements—involuntarily and unconsciously. Therefore, the term *fixational eye movements* has been introduced in the scientific literature. Microsaccades represent the fastest component of fixational eye movements and are modulated by a number of cognitive processes. After a short introduction to the physiology of the saccadic system, we described an integrative mathematical model, which is physiologically plausible and reproduces the key feature of fixational eye movements.

Decision processes in simple detection or discrimination tasks can be described by mathematical models of information accumulation via *random walks* or *diffusion processes* (Chap. 3). This model class was introduced by a mathematical psychologist. It took more than 40 years before its neurophysiological plausibility could be investigated using experimental data. With the possibility to observe neural activity via single-cell recording, it was discovered relatively recently that random walks are a surprisingly good model of the neural foundations of simple decisions.

Chapter 4 addresses the processes underlying sensorimotor integration, which is the integration of perception and motor control to generate adaptive behavior. It turned out that Bayes' rule (published 1763) is an important tool to understanding motor planning, since humans integrate their a priori knowledge with sensory perception to produce optimal behavior. We discuss a Bayesian model of eye guidance during reading that generates within-word landing positions for saccadic eye movements. However, this type of model is confined to oculomotor processes and does not address the interaction between word processing and saccadic selection, which is the topic of the next chapter.

Reading involves the coordination of several of the key cognitive subsystems, e.g., vision, attention, memory, and motor control (Chap. 5). Mathematical models of eye guidance in reading are based on a number of assumptions for the coupling of all contributing subsystems. We discuss two types of these models that make different assumptions on attention allocation—a serial-attention model (E-Z Reader, 1998) and spatially distributed attention model (SWIFT, 2002). Moreover, we show how we can use these computational models to derive quantitative hypotheses on experiments.

In Chap. 6, we discuss the more general problem of saccadic selection during natural scene viewing. The underlying model architecture is an activation-based map of attention with inhibitory map of recently fixated image patches. It turns out that *spatial point processes* provide important insights into the statistics of eye movements.

We study bimanual coordination (Chap. 7) as an example of *nonlinear coupled oscillations* that are mathematically described by differential equations. The framework proposed by Haken, Kelso, and Bunz (1986) was one of the paradigmatic models of paving the way for dynamical cognitive psychology.

The final chapter discusses and compares the models discussed before and works out typical challenges in cognitive modeling. All chapters close with exercises and comments on the literature for further reading. Advanced sections are indicated by an asterisk (*) and can be skipped during first reading. Computer code[1]in the *R Project for Statistical Computing* is available as online material via *Open Science Framework*.

Potsdam, Germany
September 2020

Ralf Engbert

[1] Visit the repository at https://osf.io/y3khb/.

Acknowledgments

This textbook has been written over several years during teaching dynamical cognitive modeling to students of Cognitive Science, Psychology, and Computer Science at the University of Potsdam. My special thanks go to the students of the summer semesters from 2017 to 2019 for their comments that helped to improve the book, especially to Lisa Schwetlick for intensive proofreading. I would also like to thank the colleagues of the Collaborative Research Centers 1294 *Data Assimilation* and 1287 *Variability in Language*, funded by the Deutsche Forschungsgemeinschaft, for the intensive scientific exchange on modeling cognitive processes. I am particularly thankful to Reinhold Kliegl, Felix Wichmann, Sebastian Reich, Shravan Vasishth, and Wilhelm Huisinga.

Contents

Chapter 1
Neural Coding

Nerve cells or neurons are specialized cells of the human body that process electrical signals and transmit output to other cells. The most important signals to neural functioning are *action potentials* which produce a short-lived and strong change of a cell's membrane potential. In the first chapter of the lecture, we discuss the analysis of series of action potentials. For example, visual signals induce neural responses that can be described by a time-dependent firing rate. A sensory neuron's activity can be modulated by presenting characteristic stimuli within its *receptive field* that codes stimulus properties of a small part of the environment by a time-dependent probability of generating action potentials. The relation between sensory stimuli and resulting neural activity is a central topic in *computational neuroscience*. In this chapter, we follow the discussion presented in Dayan and Abbott's book [3].

1.1 Properties of Neurons

Nerve cells are the basic elements of the complicated signal processing system, the nervous system, which animals need for the generation of adaptive behavior. Using cell staining techniques, it is possible to highlight a small fraction of neurons within the nervous tissue (Fig. 1.1); in addition, wiring structures between neurons are made visible qualitatively.

Nerve cells have extensive branches called neurites that can be further distinguished into *dendrites* and *axons*. This specialized cell morphology enables the signal transmission from one neuron to its neighboring cells, where a single neuron can make contact to up to 100,000 other neurons. The typical signal transmission within a neuron starts from the dendrite (input zone) to the axon (output zone).

Short axons extend from one neuron to its neighbors; others can be long enough to cover parts of the central nervous system, e.g., connections between the cortical hemispheres. In the gaps between neurons, the *synapses*, the interactions between

R. Engbert, *Dynamical Models In Neurocognitive Psychology*,
https://doi.org/10.1007/978-3-030-67299-7_1

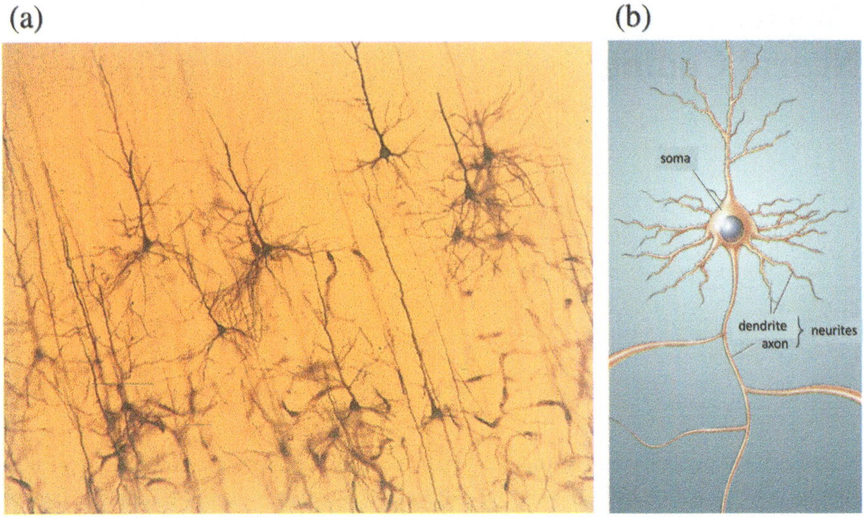

Fig. 1.1 Structures of neurons (from: [1]). (**a**) In neural tissue, a few percent of nerve cells are made visible by staining techniques using impregnation with silver salts. (**b**) Schematic illustration of a cortical pyramidal cell. Electrical signals are received at the dendrite, processed in the soma, and transmitted to a neighboring cell along the axon. Dendrites and axons are forms of branching structures called neurites

neurons are typically based on chemical signals (neurotransmitters), however, electrical synapses are also found in the nervous system of many species. In cortical cells of the mouse, axons have a typical length of 40 mm, while dendrites show an overall length of 10 mm. Counting the number of synaptic connections per unit length along the neurites gives a numerical estimate of about 2 synapses per μm [3].

Electrical potentials are generated by an imbalance of ions between intracellular and extracellular fluids. The cell membrane is found to be semipermeable due to ion channels and transporters that regulate ion gradients. Ion channels are transmembrane proteins that control the selective flow of ions across the membrane. Given the equilibrium distribution of typical ions (most important are Na^+, K^+, Ca^{2+} und Cl^-), we can calculate the membrane potential at rest, called the equilibrium potential, using the Nernst equation (if a single ion type is involved) or the Goldman equation (for several ion types) [3]. While the resting membrane potential is typically negative, the membrane potential can become positive during very short intervals (on the order of milliseconds). This process is termed *action potential* (Fig. 1.2a). The action potential is initiated by a transient increase in membrane conductance for Na^+ ions (Fig. 1.2b). For typical neurons, there is a non-zero baseline rate at which action potentials are generated at rest.

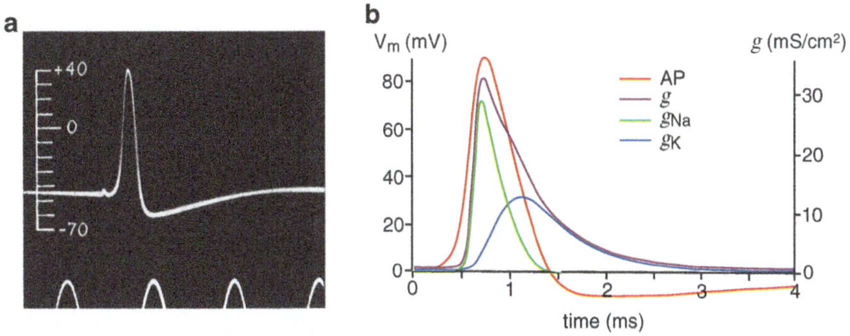

Fig. 1.2 Action potential and membrane permeability. (**a**) First measurement of a neuron's action potential [9]. (**b**) Underlying temporal variation of the membrane permeability for Na^+ and K^+ ions (from: [7])

1.2 The Early Visual System

If a neuron is stimulated by an appropriate stimulus, then the firing statistics of the neuron are modulated. In this section, we review some basic facts on the early visual system [11, 12, 14], i.e., how visual neurons are stimulated by visual input. The neural pathway of the early visual system starts with the structures of the photoreceptors of the retina in the eye. In the output layer of the retina, ganglion cells project onto the lateral geniculate nucleus (LGN), a relay structure that is part of the thalamus controlling the inflow of almost all sensory information. From the LGN cells, axons transmit activation to the primary visual cortex (Fig. 1.3) in the brain's occipital lobe. For humans and other primates with their eyes in frontal position, potentially overlapping information from the left and right parts of the visual fields must be merged in a meaningful way. Axons from retinal cells cross each other at the *optic chiasm*, so that left and right visual field are processed separately in the two cerebral hemispheres (red and blue schematic neural pathways in Fig. 1.3).

For any sensory system, the generation of neural activation from a physical stimulus is the critical process known as *signal transduction*. Based on signal transduction, changes in the environment irritate our nervous system, which then constructs our perception from incoming stimulation [14]. The process of signal transduction in sensory neurons represents the system boundary between environment and organism. Each sensory neuron codes specific properties of a given stimulus by its neural activity.

Signal transduction in the photoreceptors of the retina transforms electromagnetic radiation of the visual part of the spectrum (wavelengths between 400 nm and 700 nm) into changes of electrical membrane potentials. Two types of photoreceptors can be distinguished. *Rods* are specialized for vision under low luminance conditions, while *cones* are optimized for high-acuity color vision and the detailed analysis of stimulus properties under high luminance conditions. Cones are highly

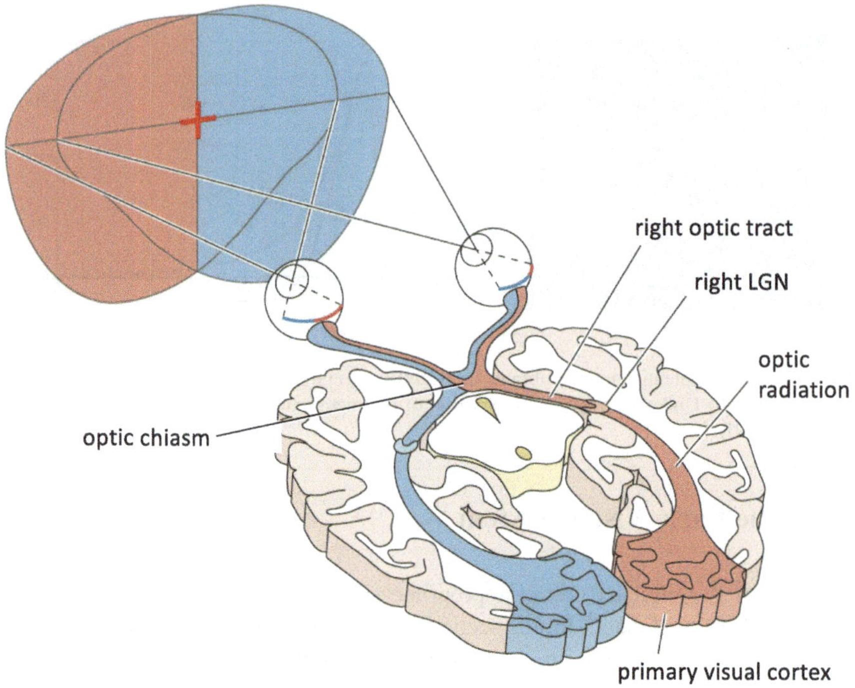

Fig. 1.3 Schematic illustration of the pathways of early visual processing. After signal transduction in the photoreceptor cells of the eyes, action potentials are generated in retinal ganglion cells. Axons from both eyes cross at the *optic chiasm*. Cells from the lateral retina project to the ipsilateral visual cortex, while cells from the nasal retina project to the visual cortex in the contralateral hemisphere (from: [1])

concentrated in a very small central portion of the retina (the fovea). The highest concentration of rods can be found in the peripheral parts of the retina.

In both types of photoreceptors, increase of luminance induces a modulation of the electrical membrane potential. The changes of the membrane potentials are transmitted to the ganglion cells (output cells) via the so-called bipolar cells. Ganglion cells are the first cells along the visual pathway that generate action potentials (Fig. 1.4). With the help of action potentials, signals can be transmitted without loss along the optic nerve that is made from the ganglion cells' axons to the lateral geniculate nucleus. Lateral connections with horizontal cells combine the input from several photoreceptors. Thus, the activity of ganglion cells represents preprocessed signals that emerge from the neural responses of many photoreceptors.

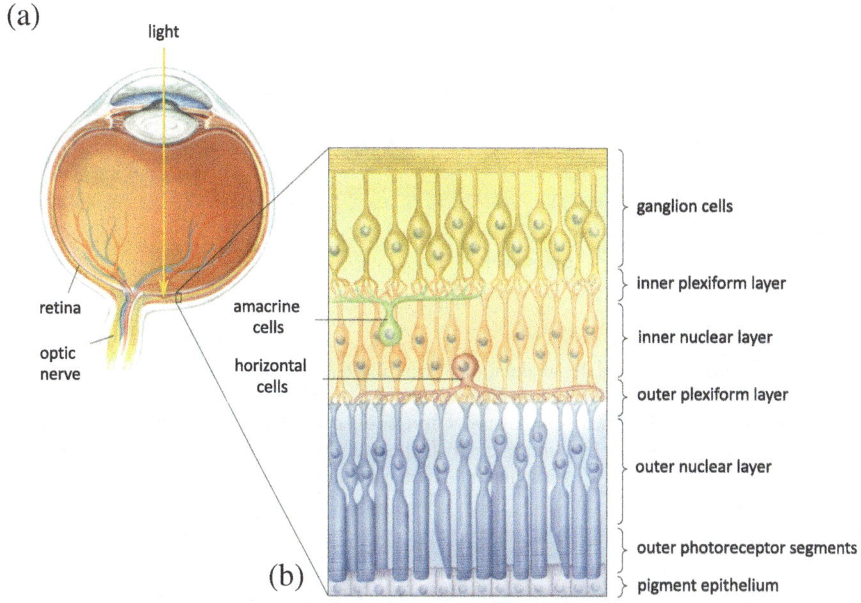

Fig. 1.4 Eye and structure of the retina (from: [1]). (**a**) Position of the retina within the eye ball. (**b**) Structure of the retina. Photoreceptors (bottom) are responsible for signal transduction. Horizontal cells provide lateral connections that are needed for the construction of center-surround receptive fields in the ganglion cells (top)

1.3 Receptive Fields

Neurons in the retina, in the LGN, and in the primary visual cortex respond only to special types of stimuli if presented in a relatively small region of the visual field. This sensitive region is defined as the neuron's *receptive field*. The spatial extent of the receptive field varies from about $0.1°$ visual angle in the fovea to several degrees in the periphery.

In a typical ganglion cell, the structure of the receptive field is divided into center and surround, since different activity patterns result from the stimulation of these two parts of the receptive field. If a spot of light is used to stimulate the center of an *OFF–center* cell, the neural firing rate decreases; if a dark spot is presented to the OFF-center, the firing rate increases transiently (Fig. 1.5b). In *ON–center* cells, the induced activity patterns are opposite to the patterns seen in OFF–center cells, i.e., stimulation of the excitatory center produces an increase of the firing rate. An important consequence of these types of receptive fields is that corresponding neurons are optimal for the detection of contrast edges in luminance, while neural activity is relatively insensitive against diffuse illumination (Fig. 1.5a).

Let us consider a stationary (i.e., time-independent) signal $S(x, y)$ over the 2D space with coordinates x and y. Mathematically, the relation between input $S(x, y)$ and output O for a neuron can be written as

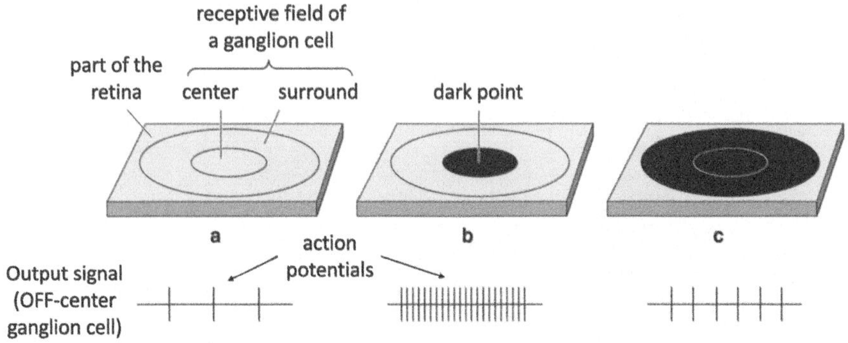

Fig. 1.5 Receptive fields and the modulation of neural activity by visual stimulation. In receptive fields with *center–surround* structure firing rates show a strong dependence on stimulus position within the receptive field. The example shows an OFF-center ganglion cell. (**a**) At rest (diffuse stimulation), the cell fires at baseline rate. (**b**) If the inhibitory center is no longer illuminated, then the firing rate decreases. (**c**) Without stimulation (no stimulus), the firing rate is slightly above baseline (from: [1])

$$O = \int_x \int_y \mathrm{d}x\, \mathrm{d}y\, K(x, y)\, S(x, y) + \xi \,, \tag{1.1}$$

where $K(x, y)$ is the receptive field of the neuron and ξ represents the spontaneous firing rate that is related to resting activity [15]. The receptive field of a ganglion cell with the typical center-surround spatial structure is often modeled as a difference of two Gaussians [6],

$$K(x, y) = \frac{w_c}{\sigma_c^2} \exp\left(-\frac{x^2 + y^2}{2\sigma_c^2}\right) - \frac{w_s}{\sigma_s^2} \exp\left(-\frac{x^2 + y^2}{2\sigma_s^2}\right) \,, \tag{1.2}$$

where the strengths are w_c for the center and w_s for the surround and the standard deviation for the center is smaller than the standard deviation for the surround, $\sigma_c < \sigma_s$, while the two Gaussians are centered at the same location. The firing rate r of the neuron is a function of the neural output O activation, i.e., $r = g(O)$.

When investigating the properties of neurons along the visual pathway, we observe that the complexity and size of the corresponding receptive fields increase as we progress to cortical areas. As an example, we find cells with orientation-selective receptive fields in the primary visual cortex. In Fig. 1.6, it is illustrated schematically, how an orientation-selective cells can be constructed from the combination of the outputs of simple cells with center-surround receptive fields. Here, ganglion cells form a linear array, so that a postsynaptic cortical cell produces high firing rates due to excitatory connections, when all ganglion cells fire simultaneously. As a result of this wiring of neural activity, the cortical cell has a linearly extended receptive field with an excitatory center that is maximally stimulated, if a bar-like stimulus of the preferred orientation is presented. An inhibitory surround is found for such an orientation-selective cell.

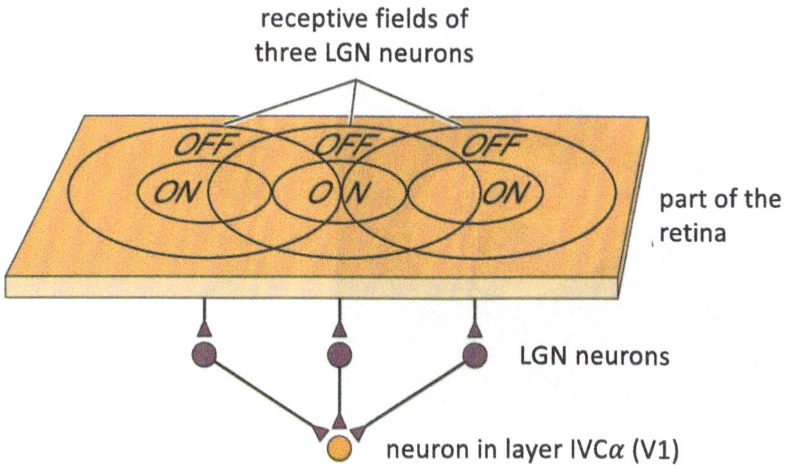

Fig. 1.6 Construction of orientation-selective receptive fields from center-surround cells (from: [1])

The center-surround structure of the receptive field is found in cells of the retina, the LGN, and the primary visual cortex and has been termed *simple cell* in the cortex. To study the properties of simple cells, their cell activity is recorded using an extremely small electrode, while stimuli are presented under variation of the orientation. An example for the dependence of a simple cell's firing rate is illustrated in Fig. 1.7.

1.4 Neural Activity as a Point Process

Action potentials are the key signals of information transmission between neurons. The specific form of an action potential or *spike* does not vary strongly over time and is less important than the exact point-of-occurrence in time. Thus, for many scientific questions, measurements of electrical activity of neurons can be reduced to the sequence of onset times, i.e., the point in time where the neuron *fires*. The resulting time series can be interpreted as a realization of the *stochastic point process* that is fully described by the onset times.

Not only spikes of electrical activity but also human behavior can often be described as a point process (see Chap. 2). Eye movements are ideal for observing neural processes in real time via monitoring of behavior [8]. Moreover, the eye is part of both sensory and motor systems and provides an ideal interface for studying sensory-motor interaction. Even when we inspect a motionless target object, our eyes perform miniature movements which occur unconsciously and involuntar-

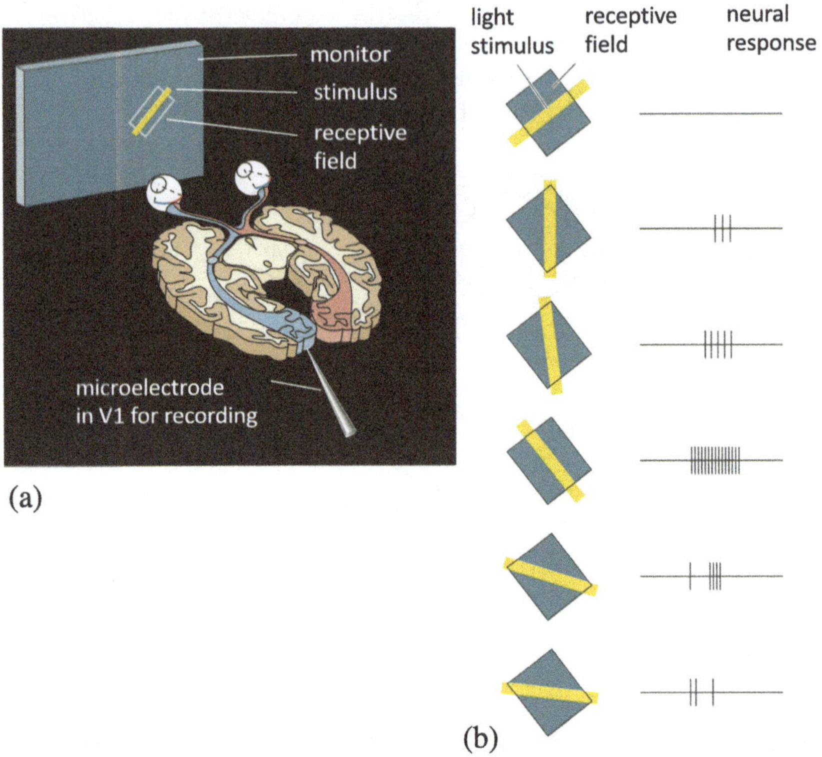

Fig. 1.7 Dependence of a simple cell (V1) on stimulus orientation (from: [1]). (**a**) Experimentally, a bar-like light stimulus is moved over the receptive field of the neuron and the firing rate are recorded using a microelectrode in V1. (**b**) The neural responses (impulses per time unit) indicate a maximum rate at an orientation of $-45°$ visual angle

ily. Here we focus on rapid, small-amplitude movements called *microsaccades* (Fig. 1.8). The durations of microsaccades (10–40 ms) are small compared to the typical interevent times (200–300 ms). Therefore, microsaccades are qualitatively very similar to neural action potentials, so that methods for the statistical analysis of neural spike trains can be used to investigate microsaccades [4].

The occurrence of action potentials (or microsaccades) at N points over time in the sequence $t_1, t_2, t_3, \ldots, t_N$ can be described mathematically as a summation over infinitesimal, idealized *spikes* [3], where the height of the spikes is not limited. The series of spikes (*spike train*) is mathematically expressed in the form of a sum of Dirac's δ-functions, i.e.,

$$\rho(t) = \sum_{i=1}^{N} \delta(t - t_i) , \tag{1.3}$$

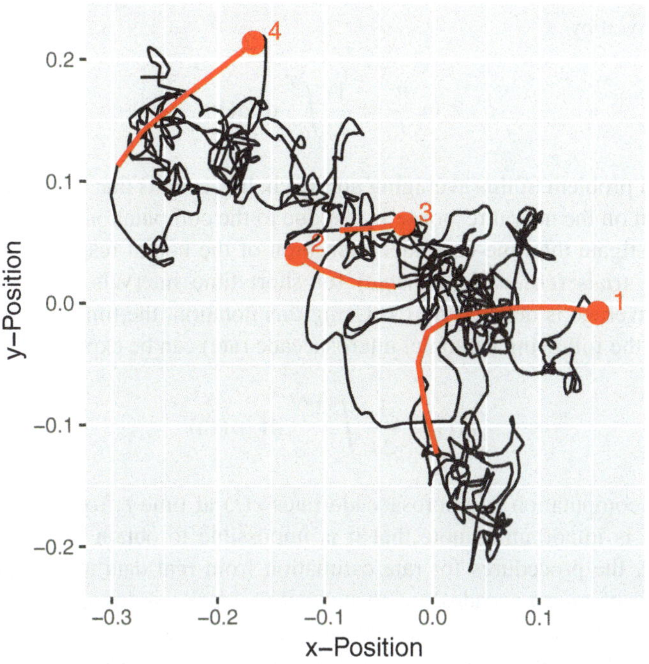

Fig. 1.8 Microsaccades (red color) are rapid, small-amplitude movements that are produced when a human observer inspects a stationary target object. Red numbers are related to endpoints of microsaccades. Microsaccades are embedded into slower random movements (black color)

where the δ-function has the properties

$$\int \mathrm{d}t\delta(t) = 1 \qquad \text{and} \qquad \int \mathrm{d}t'\delta(t-t')f(t') = f(t)\,. \tag{1.4}$$

Using this definition of the *neural response function* $\rho(t)$, we can write the sum of values of an arbitrary, well-defined function $h(t)$ at times t_1, t_2, ... as an integral over time, i.e.,

$$\sum_{i=1}^{n} h(t-t_i) = \int_{-\infty}^{\infty} \mathrm{d}\tau h(\tau)\rho(t-\tau)\,. \tag{1.5}$$

Since neurons are very different from the deterministic elements we use in electronic circuits, the sequence of action potentials from a neuron varies randomly from one realization to the next realization, even if the input is exactly the same. A straightforward approach to the statistical analyses of spike trains is to compute temporal averages over a certain time interval $(0,\ T)$. Mathematically, this temporal average equals the integral of the neural response function over this time interval, divided by the length T of the interval. Therefore, the firing rate in time interval $(0, T)$ is given by

$$r = \frac{n}{T} = \frac{1}{T}\int_0^T d\tau \rho(\tau) \,. \tag{1.6}$$

The central problem of this averaging approach, however, is that all time-dependent information on the neural response is lost due to the computation of the average.

To investigate the time-dependent variation of the neural response, we average over many trials (*ensemble average*) for short time intervals $(t;\, t + \Delta t)$. The ensemble average is denoted by $\langle . \rangle$. Using this notation, the time-dependent firing rate (or, in the following example, microsaccade rate) can be expressed as

$$r(t) = \frac{1}{\Delta t}\int_t^{t+\Delta t} d\tau \langle \rho(\tau) \rangle \,. \tag{1.7}$$

For the computation of microsaccade rates $r(t)$ at time t from a finite number of trials, it is important to note that it is impossible to obtain an exact estimate. In Fig. 1.9, the procedures for rate estimation from real data are illustrated. The onset times of microsaccades are visualized as small circles in Fig. 1.9a, where each horizontal line represents an experimental trial and different colors represent observations from different observers. The microsaccade rate is computed from these raw data by weighted averaging using a window function $w(t)$,

$$r_{approx}(t) = \sum_{i=1}^{n} w(t - t_i) \,, \tag{1.8}$$

where, in the simplest case, we use a rectangular window function defined by

$$w(t) = \begin{cases} 1/\Delta t : \text{if} & -\Delta t/2 \le t < \Delta t/2 \\ 0 & : \text{otherwise} \end{cases} \,. \tag{1.9}$$

Now, given the above definition of the neural response function $\rho(t)$, we can rewrite the sum as an integral over the product of the window function and the neural response function, i.e.,

$$r_{approx}(t) = \int_{-\infty}^{\infty} d\tau w(\tau)\rho(t - \tau) \,, \tag{1.10}$$

where the integral is called a linear filter and the window function is denoted as a (filter) *kernel*. The concept of a filter kernel provides a flexible tool to adapt the properties of the kernel to the boundary conditions from the data and to the needs arising from the scientific question to be studied. As an example, we can use a Gaussian function as a kernel.

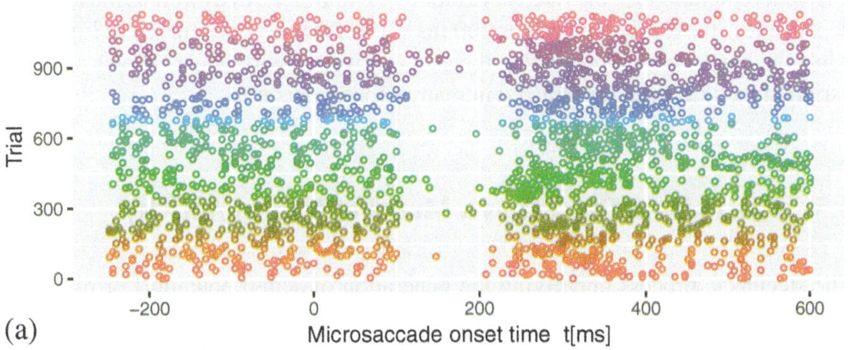

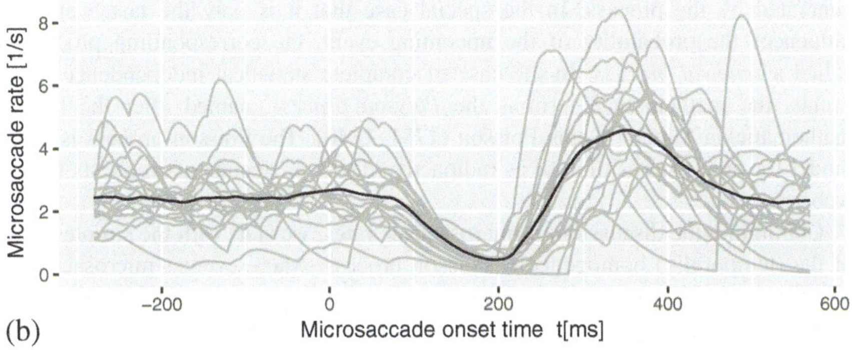

Fig. 1.9 Estimation of the time-dependent variation of microsaccade rate relative to presentation of a visual stimulus. The stimulus appeared on a computer display at time $t = 0$ ms, while a human observer was required to look at a fixation cross and high-precision measurements of eye movements were made. (**a**) Each dot indicates the onset of a microsaccade where the color indicates the participant. The data shown are taken from $N = 1128$ trials from 20 human observers. (**b**) Estimated microsaccade rates from the data above using the causal filter. Individual curves for the modulation of microsaccade rate for several participants (gray lines) and average over 20 participants (black line)

An important window function that is frequently used in the estimation of firing rates from neural spike trains has the property that the estimate at time t cannot be biased from spike events that occur later in time. Such a *causal* kernel can be defined as

$$w(\tau) = \left[\alpha^2 \tau \exp(-\alpha\tau)\right]_+ , \tag{1.11}$$

where the $[.]_+$ operation takes only the positive part of the argument and vanishes elsewhere. In Fig. 1.9b, the resulting time-dependent estimate of the microsaccade rate is plotted for a causal kernel with parameter $\alpha = 1/20$.

After first analyses of stimulus-induced modulation of microsaccade rates [5], it became immediately clear that there is large interindividual variation for stationary microsaccade rates (see the rate curves for $t < 0$ in Fig. 1.9b) as well as the temporal dependence (most pronounced for 200 ms$< t <$ 500 ms; gray curves in Fig. 1.9b), while the black curve representing the average over all participants shows a smooth variation that has been reproduced in many experiments now.

1.5 The Poisson Process as a Model of Neural Activity

The stochastic process underlying the generation of action potentials or microsaccades (as two examples for neural activity) is called a *point process* [2]. In such a process, the probability of an event can be depending on all earlier events generated by the process. In the special case that it is only the last event that influences the probability of the upcoming event, the corresponding process is called a *renewal process*. In the case of complete statistical independence of all events, the mathematical term is the *Poisson process*, named after the French mathematician Siméon Denis Poisson (1781–1840). The Poisson process is a key model for processes as diverse as radioactive decay or documents requests from a web server.

Continuing the discussion of microsaccade rates, we start with the simplest case of the (temporally) homogeneous Poisson process, where events (microsaccades) are generated with constant rate $r(t) \equiv r$ over time t. Suppose that we observe n events at times $t_1 < t_2 < \ldots < t_n$ during an observation window of length T. The probability $P_T[n]$ that exactly n microsaccades are observed in the interval $(0; T)$ can be calculated as a product of three terms.

To derive an equation for the probability, we subdivide the interval of length T into M smaller intervals (*bins*) of length $\Delta t = T/M$. Here we assume that the duration Δt is small enough that not more than a single event can be observed per bin. Practically, this can always be achieved, since Δt can be made arbitrarily small. The probability that one event (microsaccade) is observed in a bin of length $\Delta t = T/M$ is given by $r\Delta t$, since r is the (constant) rate. Therefore, the probability of observing one event in each of n bins is $(r\Delta t)^n$. Moreover, the probability of finding no event in the remaining $M-n$ bins is $(1-r\Delta t)^{M-n}$. Finally, the binomial coefficient $M!/((M-n)!n!)$ gives the number of combinations of distributing n events over M bins. This factor is necessary, since we cannot distinguish the possible combinations of events over bins. The product of the three factors gives, in the limit of $\Delta t \to 0$, an equation for the probability of observing n statistically independent events during the interval of length T, i.e.,

$$P_T(n) = \lim_{\Delta t \to 0} \frac{M!}{(M-n)!n!}(r\Delta t)^n(1-r\Delta t)^{M-n} \ . \tag{1.12}$$

For the calculation of the limit $\Delta t \to 0$, we note that the number of bins, M, will become infinitely large, since $M = T/\Delta t$. Therefore, the approximation $M - n \approx M$ is valid in the limit. Using the definition $\epsilon := r\Delta t$, the last term of Eq. (1.12) can be written as

$$\lim_{\Delta t \to 0}(1 - r\Delta t)^{M-n} = \left(\lim_{\epsilon \to 0}(1-\epsilon)^{1/\epsilon}\right)^{rT} = \left(\frac{1}{e}\right)^{rT} = e^{-rT} \ . \tag{1.13}$$

Furthermore, for large M we obtain $M!/(M-n)! \approx M^n = (T/\Delta t)^n$. Consequently, the probability $P_T(n)$ can be calculated as

$$P_T(n) = \frac{(rT)^n}{n!}\exp(-rT) \ . \tag{1.14}$$

This famous probability distribution is known as the Poisson distribution.

What is the corresponding statistical distribution of waiting times between two subsequent events of a Poisson process? Let us assume that the ith event was observed at time t_i. The probability of observing the next event $i+1$ at time t_{i+1} after waiting time τ in the interval $t_i + \tau \leq t_{i+1} < t_i + \tau + \Delta t$ (for short intervals Δt) can be calculated as the product of two probabilities. The first factor is the probability that *no* event is observed in the interval $(0; \tau)$, which is given by $P_\tau(n=0) = \exp(-r\tau)$. The second factor is the probability that one event is observed in following interval $(\tau; \tau + \Delta t)$, which is $r\Delta t$. Thus, the probability of observing the waiting time τ is given by

$$P\{\tau \leq t_{i+1} - t_i < \tau + \Delta t\} = r\Delta t \ \exp(-r\tau) \ . \tag{1.15}$$

The corresponding probability density of the waiting time can be written as

$$P(\tau) = r\exp(-r\tau) \ . \tag{1.16}$$

For the numerical simulation of the Poisson process, Eq. (1.16) can be exploited to construct an iterative algorithm. A stochastic sequence of onset times is generated by the iterative rule

$$t_{i+1} = t_i - \log(\xi_i)/r \ , \tag{1.17}$$

where we used the fact that the negative logarithm $-\log(\xi)$ generates an exponentially distributed random variable for an equally distributed (computer-generated) random variable ξ on the interval $(0; 1]$. This algorithm can be applied to the simple case of the homogeneous Poisson process with a constant rate.

For the inhomogeneous Poisson process with a time-dependent rate we can use a *rejection method* in combination with the iterative rule in Eq. (1.17). In the resulting algorithm, we generate a sequence for the homogeneous Poisson process and apply a thinning procedure where events are removed in order to obtain the required rate modulation [3]. First, we choose an upper bound r_{max} of the rate of the process, so that $r(t) \leq r_{max}$ for all $t > 0$. Second, we simulate a realization of onset times according to Eq. (1.17), i.e., $t_{i+1} = t_i - \log(\xi_i)/r_{max}$. Third, we apply the thinning procedure to the resulting sequence with rate r_{max}. In this type of rejection algorithm, an event i at time t_i will be removed from the sequence, if

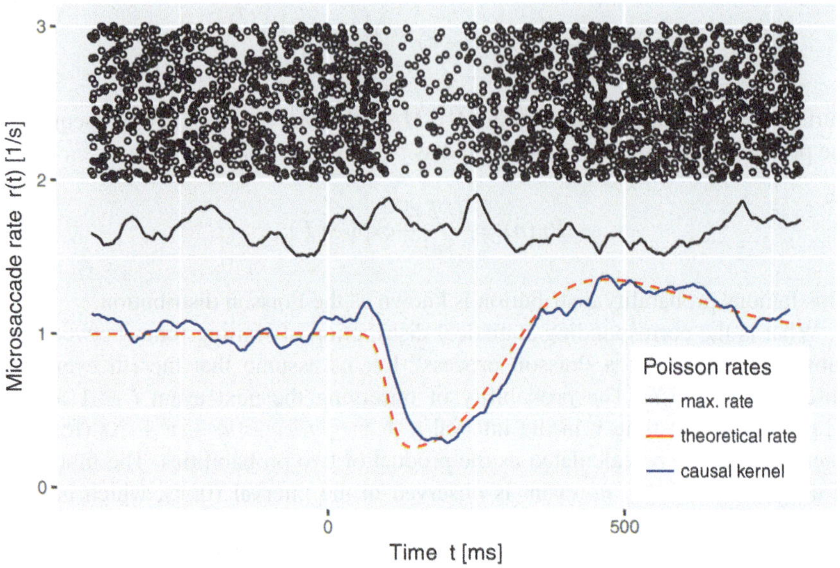

Fig. 1.10 Simulation of the inhomogeneous Poisson process. *Top panel.* Each dot represents the (simulated) onset of a microsaccade, where each row is a trial. *Bottom panel.* Estimated microsaccade rates (theoretical rate: red dashed line; maximum rate before thinning: black line; rate estimated after thinning: blue line)

$r(t_i)/r_{est}(t_i) < \chi_i$, where χ is an equally distributed random number from (0; 1). In the case $r(t_i)/r_{est}(t_i) \geq \chi_i$, we keep event i in the sequence. An example for the simulation is illustrated in Fig. 1.10, which is the topic of Problem 1.2.

In the next chapter, we will discuss models for the generation of fixational eye movements and microsaccades. The most advanced models can reproduce the rate modulation of microsaccades based on biologically plausible assumptions, linking cognitive processes to basic oculomotor statistics.

1.6 Exercises

Exercise 1.1 (Receptive Fields as Difference of Gaussians) In one dimension, the receptive field with on-center and off-surround can be written as [15]

$$K(x) = \frac{w_c}{\sigma_c} \exp\left(-\frac{x^2}{2\sigma_c^2}\right) - \frac{w_s}{\sigma_s} \exp\left(-\frac{x^2}{2\sigma_s^2}\right) . \tag{1.18}$$

Find appropriate values for the parameters w_c, w_s, σ_c, σ_s and plot the resulting curve for a receptive field which is excitatory between about $-2°$ and $+2°$ of visual angle and inhibitory outside this region. What are the corresponding parameter values?

Exercise 1.2 (Estimation of the Temporal Modulation of Microsaccade Rates) (a) Plot the microsaccade data given in file `micro.dat`. Use a small dot for each microsaccade, where the onset time is the abscissa and the trial number is the ordinate (see Fig. 1.9). (b) Use non-overlapping rectangular windows, Gaussian, and causal kernels to estimate the temporal modulation of microsaccade rates from data and compare the results via plotting.

Exercise 1.3 (Normal (Gaussian) Probability Density)

(a) Generate $N = 1000$ normally distributed random numbers with mean $\mu = 5.7$ and standard deviation $\sigma = 3.9$ using `rnorm()`.
(b) Plot a histogram using R's basic plotting command.
(c) Add a curve of the theoretically expected density to the same plot using the Gaussian

$$p(y) = \frac{1}{\sqrt{2\pi}\sigma} \exp\left\{-\frac{(x-\mu)^2}{2\sigma^2}\right\} \tag{1.19}$$

and the R-function `dnorm()`.
(d) With package `ggplot2` for plotting, repeat tasks (b) and (c) using commands `geom_histogram()`, `geom_line()`, and `geom_density()`.

Exercise 1.4 (Simulation of the Inhomogeneous Poisson Process) (a) Find a theoretical function for an idealized description of the mean rate modulation (Fig. 1.9b) and plot the corresponding function. (b) Simulate an inhomogeneous Poisson process for the rate modulation found in (a). Generate data from a corresponding homogeneous Poisson process with rate r_{max}. Apply the rejection algorithm for thinning of the homogeneous sequence to obtain the correct temporal variation of the microsaccades rate (Fig. 1.10).

1.7 Further Reading

While the generation of electrical activity in nerve cells is beyond the scope of neurocognitive modeling, there are interesting mathematical links between single-cell models and the behavioral models that represent the focus of this textbook. Moreover, neurophysiological plausibility is an important boundary condition for cognitive models. Therefore, the development and analysis of cognitive models will benefit from basic knowledge on models of neural activity. An introduction to the problems and mathematical methods of *computational neuroscience* can be found in the textbook by Dayan and Abbott [3].

Since modeling of eye movements requires knowledge of the visual system, it might be useful to consult introductions to the visual system which can be found in the textbooks by Bear et al. [1], Purves et al. [12], and Wolfe et al. [14]. See also the review article on parallel processing in the visual system [10].

Stochastic point processes represent an important mathematical framework for the analysis of neural and behavioral data. For a review on probability the reader might want to consult Appendix A.1 or the textbooks by Brémaud [2] or Taylor and Karlin [13]. More specifically, the application of Poisson statistics to the analysis of microsaccade sequences is discussed in the review paper of Ref. [4].

References

1. Bear, M. F., Connors, B. W., & Paradiso, M. A. (2018). *Neurowissenschaften: Ein grundlegendes Lehrbuch für Biologie, Medizin und Psychologie*. Springer-Verlag. https://link.springer.com/book/10.1007/978-3-662-57263-4.
2. Brémaud, P. (1999). *Markov chains: Gibbs fields, Monte Carlo simulation, and queues* (vol. 31). Springer. https://www.springer.com/de/book/9780387985091#otherversion=9781441931313.
3. Dayan, P., & Abbott, L. F. (2001). *Theoretical neuroscience*. Cambridge, MA: MIT Press. https://mitpress.mit.edu/books/theoretical-neuroscience.
4. Engbert, R. (2006). Microsaccades: A microcosm for research on oculomotor control, attention, and visual perception. *Progress in Brain Research, 154*, 177–192. https://doi.org/10.1016/S0079-6123(06)54009-9.
5. Engbert, R., & Kliegl, R. (2003). Microsaccades uncover the orientation of covert attention. *Vision Research, 43*, 1035–1045. https://doi.org/10.1016/S0042-6989(03)00084-1.
6. Enroth-Cugell, C., & Robson, J. G. (1966). The contrast sensitivity of retinal ganglion cells of the cat. *The Journal of Physiology, 187*(3), 517–552. https://doi.org/10.1113/jphysiol.1966.sp008107.
7. Häusser, M. (2000). The Hodgkin-Huxley theory of the action potential. *Nature Neuroscience, 3*, 1165–1165. https://doi.org/10.1038/81426.
8. Hepp, K. (2004). Neurodynamik in Echtzeit. *Physik Journal, 3*(8-9), 55–60. https://www.pro-physik.de/details/physikjournalIssue/1089779/Issue_8_2004.html#1097371.
9. Hodgkin, A. L., & Huxley, A. F. (1939). Action potentials recorded from inside a nerve fibre. *Nature, 144*(3651), 710–711. https://doi.org/10.1038/144710a0.
10. Nassi, J. J., & Callaway, E. M. (2009). Parallel processing strategies of the primate visual system. *Nature Reviews Neuroscience, 10*(5), 360–372. https://doi.org/10.1038/nrn2619.
11. Nicholls, J. G., Martin, A. R., Wallace, B. G., & Fuchs, P. A. (2012). *From neuron to brain* (vol. 271). Sunderland/MA: Sinauer Associates Inc. https://www.sinauer.com/from-neuron-to-brain.html.
12. Purves, D., Augustine, G. J., Fitzpatrick, D., Hall, W. C., LaMantia, A.-S., & White, L. E. (2012). *Neuroscience* (5th edn.). Sunderland/MA: Sinauer Associates Inc. https://www.sinauer.com/neuroscience-770.html.
13. Taylor, H., & Karlin, S. (1999). *An introduction to stochastic modeling* (3rd edn.). San Diego, CA: Academic Press. https://www.elsevier.com/books/an-introduction-to-stochastic-modeling/pinsky/978-0-12-381416-6.
14. Wolfe, J. M., Kluender, K. R., & Levi, D. M. et al. (2015). *Sensation and perception* (4th edn.). Sunderland/MA: Sinauer Associates Inc. https://www.sinauer.com/sensation-perception-784.html.
15. Zhaoping, L. (2014). *Understanding vision: Theory, models, and data*. Oxford: Oxford University Press. https://global.oup.com/academic/product/understanding-vision-9780199564668.

Chapter 2
Fixational Eye Movements

Visual perception is fundamentally based on motion. This fact is obvious for some sensory systems in animals; for example, a resting fly is invisible to a frog. The situation rapidly changes as soon as the fly starts to move. Thus, object movements are an essential prerequisite for sensation in frogs [14]. Since the detection of motion is critical in predator-prey relationships, high sensitivity for motion perception might have been a key advantage for the evolution of visual systems. As a consequence, our visual system evolved as a fast-adapting system to constant input, so that moving objects in the environment stand out immediately, while motionless background is suppressed. Equipped with such a visual system optimized for the detection of motion and change, however, we were unable to process fine details of a completely stationary scene without active refresh of the retinal image. Thus, unexpectedly, our eye generate miniature movements, when we fixate a stationary target object. These miniature movements counteract retinal adaptation that would induce perceptual bleaching of constant input within less than about 100 ms [4].

Because of the presence of miniature eye movements during visual fixation, the term *fixational eye movements* is used to indicate that the eyes are kept close to an intended target object, while small-amplitude movements ($<1^\circ$) are produced involuntarily. Fixational eye movements fall into three different categories: The slow component is a superposition of *physiological drift* with maximum velocities of about 30 min arc s^{-1} and *tremor*, an oscillatory component in the frequency range of 30–100 Hz [15]. Tremor is neglected very often, since its movement amplitude is small and cannot be resolved using current video-based eye-tracking technology.

The fastest component of fixational eye movements is produced by *microsaccades*, the rapid, small-amplitude movements introduced in the last chapter (Fig. 1.8, p. 9), with a typical duration of 10–40 ms and movement amplitudes of less than 1°. Different from drift, microsaccades are highly correlated between the eyes—a fact that we will exploit when we develop a detection algorithm for microsaccades (Sect. 2.4). During the last 10 years it was observed that microsaccades are modu-

R. Engbert, *Dynamical Models In Neurocognitive Psychology*,
https://doi.org/10.1007/978-3-030-67299-7_2

lated by a broad range of cognitive processes and, in particular, by selective attention [8], which makes microsaccades attractive as a new research tool for monitoring cognitive processing. We start with a short introduction to the physiology of saccadic eye movements.

2.1 Physiology of the Saccadic System

The eyes are moved by six extra-ocular muscles (Fig. 2.1a). For simplicity, we focus on horizontal movements, which are controlled by lateral (lr = lateral rectus) and medial (mr=medial rectus) eye muscles. During a saccadic eye movement, the firing rate of the lateral motor neuron increases (up to 750 Hz, see upper panel in Fig. 2.1b). This phasic activity generates an acceleration toward the target position, which results in a saccadic (ballistic) eye movement.

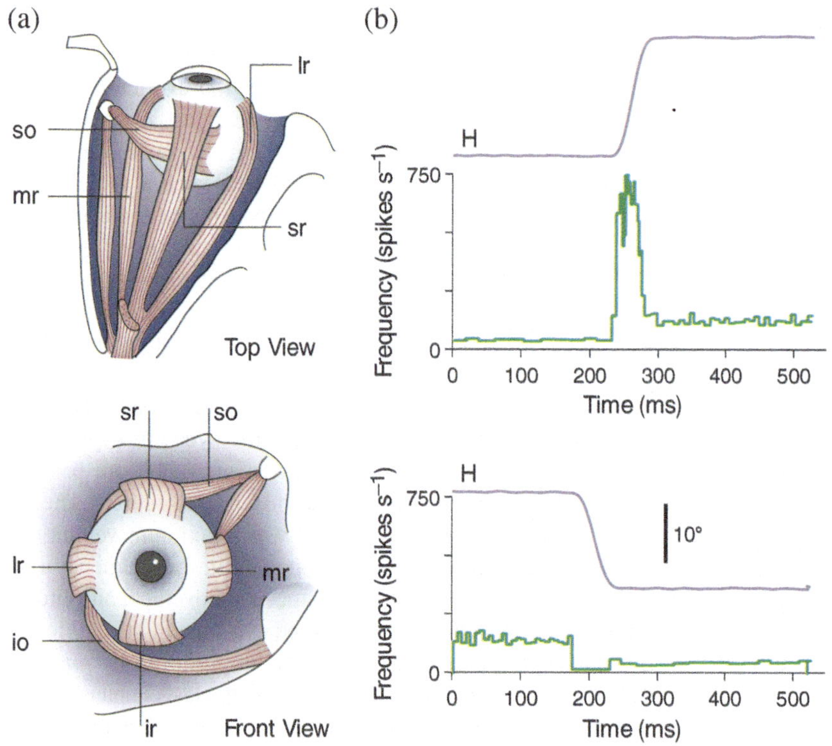

Fig. 2.1 Eye muscles and the activity of the lateral motor neuron during saccades (from: [21]). (**a**) Four extra-ocular muscles control horizontal and vertical movements, while two additional muscles induce torsional movements. (**b**) Horizontal eye movements and associated activities of the lateral motor neuron during two saccades

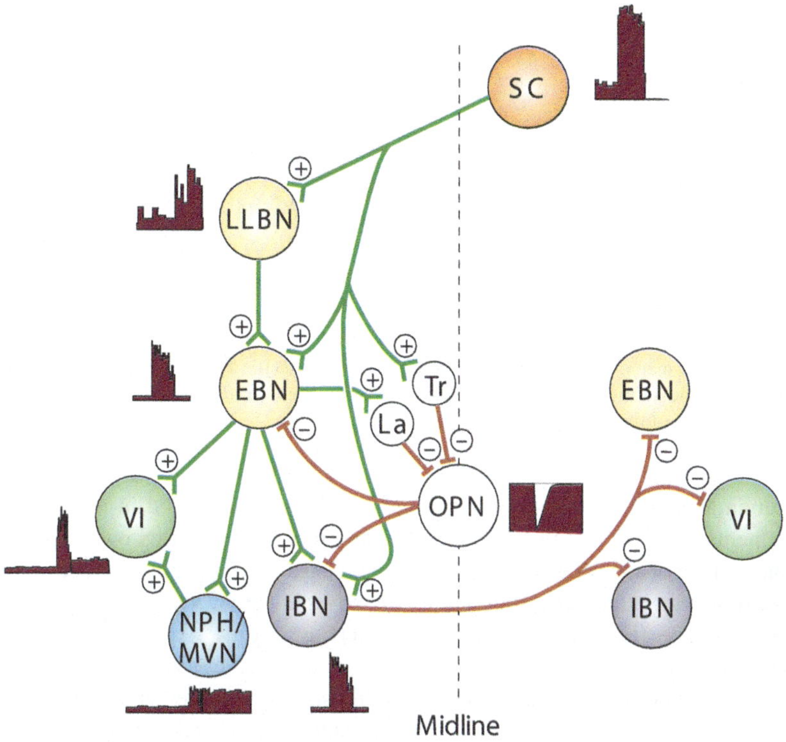

Fig. 2.2 Neuronal pathway for the generation of saccadic activity in the brainstem (from: [21]). The saccade target is represented in the neural map of the *superior colliculus* (SC), where output from descending cortical pathways is combined for saccadic selection. Two different routes of neural connections control phasic activity (via EBNs) and tonic activity (in the NPH/MVN)

After the saccade with a typical duration of about 50 ms, the firing rate of the motor neuron decreases, however, it does not relax to the baseline value, but settles to a value that is proportional to the eccentricity of the final position of the eye (tonic activity). During a saccade in the opposite direction, the lateral motor neuron is inhibited to provide for a maximum acceleration by the medial eye muscle (Fig. 2.1b, bottom panel).

The neuronal structures of the motor pathway (Fig. 2.2) generate the different components of the activity pattern during a saccadic eye movement (Fig. 2.1, right column). The phasic activity that accelerates the eye towards the saccade target is produced by *excitatory burst units* (EBN). These neurons generate a strongly increasing firing rate, while *tonic units* in the *nucleus prepositus hypoglossi* (NPH/MVN) produce the activity that is proportional to the eye position after the saccade.

The *superior colliculus* (SC) is the central structure of the brainstem control for saccadic eye movements (Fig. 2.3a). This structure receives direct input from

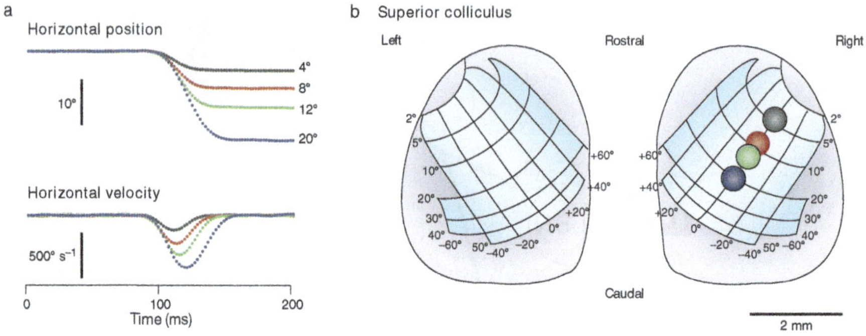

Fig. 2.3 Motor map of the superior colliculus (SC; from: [21]). (**a**) Eye position and velocity as a function of time during saccades of four different amplitudes. (**b**) Corresponding target positions in the motor map of the SC

the retina as well as from several regions of the cortex involved in the control of movements. The SC is a two-hemispheric structure, each with its own motor map for the control of saccades in the contralateral visual field. In animal experiments, it can be shown that electrical stimulation of neurons in the SC's motor layer generates saccades to a precisely defined target position (Fig. 2.3b).

The interpretation of experimental data on saccadic eye movements depends crucially on our knowledge of the neurophysiological organization of the oculomotor system. It turned out that microsaccades are basically controlled by the same neurophysiological structure as voluntary saccades that are produced during everyday activity (e.g., reading, visual search). Thus, some aspects of movement patterns during visual fixation (i.e., fixational eye movements) might be explained by the physiology of the saccadic system and, in particular, by activity patterns in the SC.

From this perspective, neural plausibility provides important constraints on model assumptions for the development of mathematical models of fixational eye movements. Before we start to discuss different models of fixational eye movements and microsaccades, we first investigate statistical properties of fixational eye movements.

2.2 Properties of Fixational Eye Movements

Because of its erratic nature, a typical trajectory of fixational eye movements can be characterized by statistical measures only. In general, different qualitative types of random motions (*random walks*) can be distinguished [17] by the probability density of movement steps (increments) and by statistical correlations between movement increments.

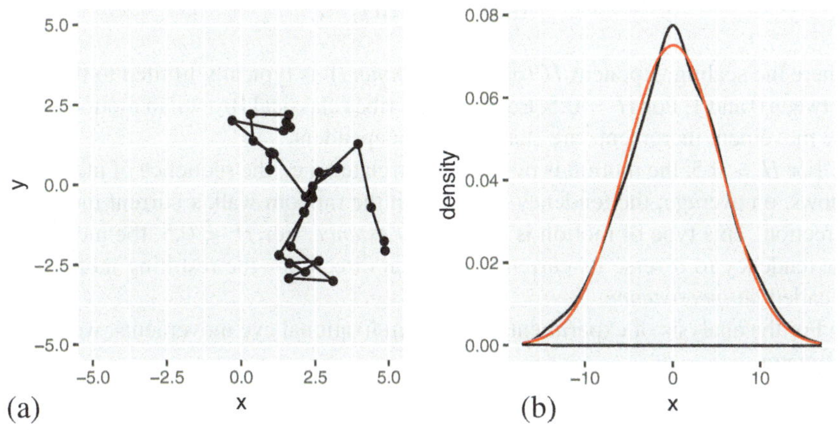

Fig. 2.4 Simulated random walk. (**a**) Random walk with normally distributed increments (30 steps). (**b**) Distribution of the x-coordinates of the endpoints from 1000 realizations

An example for a simulated random walk is given in Fig. 2.4a. In each step of the random walk, a step with length sampled from a normal distribution is performed in a random orientation with uniformly distribution turning angle. The distribution of the x-coordinate of the endpoint of this random walk after 30 iteration steps is given in Fig. 2.4b (black line). The resulting distribution is in very good agreement with a Gaussian (red line). Interestingly, the fact that the distribution is approximately Gaussian is, under some constraints, independent of the type of the distribution of the step length. This is a result of the famous *central limit theorem*: The sum of n independent random variables X_i for $i = 1, 2, \ldots, n$ is approximately Gaussian distributed for $n \to \infty$, if the variance $\mathrm{var}(X_i)$ is finite and the Lindeberg condition holds (see Appendix A.1). Due the central limit theorem, we expect to see normal distributions, whenever an observable is based on the addition of many small and independent random effects. Brownian motion is probably the most famous example [1]. In this case, a huge number of fluid molecules generate random kicks to a pollen grain suspended in the fluid. Thus, the distance of the Brownian particle travelled in a certain amount of time is normally distributed.

The correlation between movement increments can be analyzed by computing the mean square displacement $\langle(\Delta x)^2\rangle$ of the process as a function of the temporal delay Δt between two measurements (*two-point correlation function*). For Brownian motion as the most important example of a random walk, the mean square displacement grows linearly with the time interval Δt [5]. This result was derived by Einstein in 1905 based on the assumption that increments of Brownian motion are uncorrelated, since the Brownian particle is driven by statistically independent hits from fluid molecules.

A very important generalization of classical Brownian motion was introduced by random walks that are governed by a power-law relation [13] between mean square displacement and observation interval, i.e.,

$$\langle (\Delta x)^2 \rangle \propto (\Delta t)^{2H} , \tag{2.1}$$

where the scaling exponent H (or *Hurst exponent*) is typically limited to the range between 0 and 1. For $H = 0.5$, Eq. (2.1) describes classical Brownian motion, where the movement increments are statistically independent.

For $H > 0.5$, the motion is positively correlated, i.e., the sequence of increments shows, on average, the tendency to maintain the random walk's current movement direction. This type of motion is termed *persistence*. For $H < 0.5$, the motion has the tendency to reverse its current movement direction—the resulting motion type is called *antipersistence*.

For the analysis of experimental data from fixational eye movements, we use the estimator

$$D^2(m) = \frac{1}{N-m} \sum_{i=1}^{N-m} \|\mathbf{x}_{i+m} - \mathbf{x}_i\|^2 , \tag{2.2}$$

where m is the time lag that plays the role of Δt in Eq. (2.1). The Hurst exponent can be obtained as the slope of the plot of $D^2(m)$ versus m for double-logarithmic scaling. The procedure is schematically illustrated in Fig. 2.5a. If this analysis is applied to experimental data of fixational eye movements, we observe a linear relation between mean square displacement and temporal delay on two different

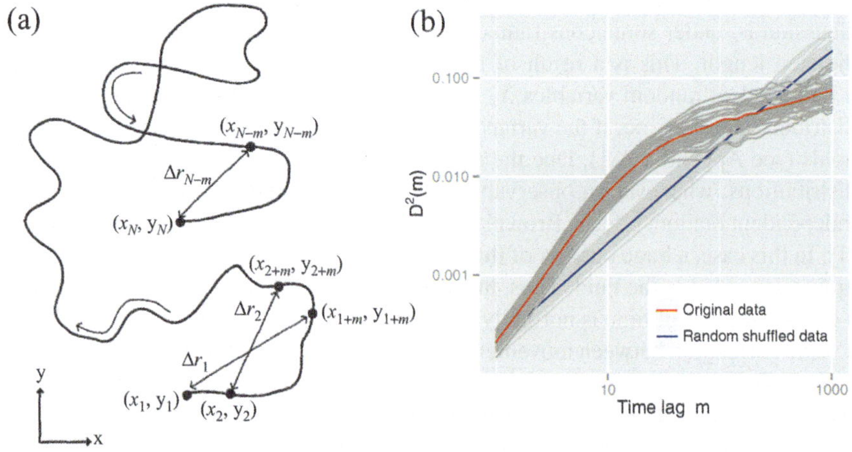

Fig. 2.5 Statistical correlations of fixational eye movements. (**a**) Illustration of the computation of the mean square displacement of two data samples at a temporal delay of m samples (from: [3]). (**b**) A plot of the mean square displacement as a function of the temporal delay indicates the existence of two different timescales (single trials: grey lines; average: red line), where a transition from persistence on the short time scale to antipersistence on the long time scale is observed (after: [6]). For randomly shuffled increments, a constant slope of one is observed (randomized single-trial data: light grey; average: blue line)

scales (Fig. 2.5b). On a short time scale ($\Delta t \leq 50$ ms), the motion is persistent, while on a longer time scale ($\Delta t \geq 100$ ms) we observe antipersistence [9].

A qualitatively very similar result was found by Collins and De Luca [3] based on center-of-pressure time series recorded from postural fluctuations during quiet standing on a force plate. Thus, statistical correlations of a similar type are characteristic for very different motor systems. The fact that some aspects of statistical behavior in movement fluctuations generalize across very different motor systems may be looked upon as a key motivation for the development of mathematical models.

2.3 The Role of Temporal Delays

In the nervous system, statistical fluctuations are present over large spatial and temporal scales, from single cells to behavior. Several mathematical models have been proposed or the transition from persistence to antipersistence of fluctuations, which was discussed in the previous section. An important approach to model building is based on explicit modeling of physiological time delays, since neural transmission times, e.g., from peripheral receptors to somatosensory cortex, are an unavoidable source of temporal delays in the interaction of neural subsystems. Therefore, random motions with time delay represent an interesting model for the generation of statistical correlations in fixational eye movements.

In an important study in 1995, Ohira and Milton [18] investigated *time-delayed random walks*. First, without time delay, the authors assumed that the movement direction of random walks, defined on a line of discrete positions $X(t) \in \{0, \pm 1, \pm 2, \ldots\}$ at discrete times $t = 0, 1, 2, \ldots$, is described by the probability $Q(t)$ to step to the right (positive direction) at time t with

$$Q(t) = \begin{cases} p & : X(t) > 0 \\ 1/2 & : X(t) = 0 \\ 1-p & : X(t) < 0 \end{cases} . \tag{2.3}$$

The resulting random walk is stable around $X = 0$, if $0 < p < 1/2$ [18]. Moreover, because of symmetry with respect to the origin at $X = 0$, the average position is $\langle X(t) \rangle = 0$.

Based on the iteration rule (Fig. 2.6) of the random walk without delay, Eq. (2.3), it is possible to calculate the stationary probability distribution. First, we exploit the symmetry of the problem and derive the following set of equations for the probability $P_X(t)$ that the walker is at position X at time t,

$$P_0(t+1) = (1-p)P_1(t) + (1-p)P_{-1}(t) = 2(1-p)P_1(t) \tag{2.4}$$

$$P_1(t+1) = \frac{1}{2}P_0(t) + (1-p)P_2(t) \tag{2.5}$$

$$P_X(t+1) = pP_{X-1}(t) + (1-p)P_{X+1}(t) \quad \text{for} \quad 2 \leq X . \tag{2.6}$$

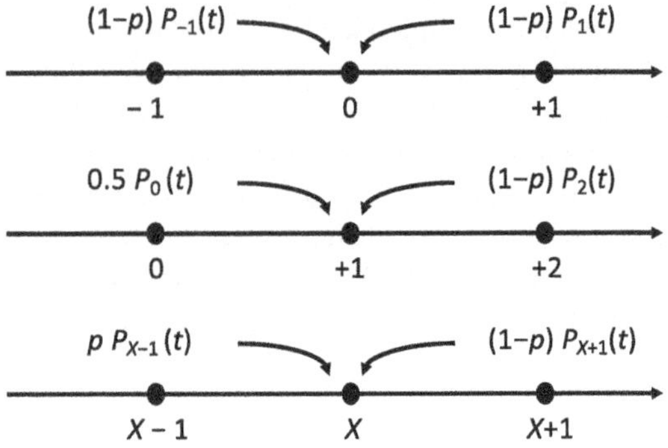

Fig. 2.6 Derivation of evolution equations, Eq. (2.4), for the Ohira-Milton model of time-delayed random walks

Now we assume that a stationary (i.e., time-independent) solution for the probability $P_X(t) = P_X^s = \text{const}$ exists. The solution can be calculated by choosing the ansatz

$$P_X^s = C_1 + C_0 \left(\frac{p}{1-p}\right)^X \qquad (1 \le X) , \tag{2.7}$$

where C_0 and C_1 are constant parameters. Because of the normalization of probability, $\sum_{X=-\infty}^{+\infty} P_X^s$, we obtain $C_1 = 0$. From Eq. (2.4) it follows that

$$P_0^s = 2(1-p)P_1^s = 2(1-p)C_0 \frac{p}{1-p} = 2pC_0 . \tag{2.8}$$

Moreover, from normalization and the ansatz, Eq. (2.7), we obtain

$$1 = \sum_{X=-\infty}^{+\infty} P_X^s = P_0^s + 2 \cdot \sum_{X=1}^{+\infty} P_X^s = 2pC_0 + 2C_0 \left(\sum_{X=0}^{+\infty} \left(\frac{p}{1-p}\right)^X - 1 \right) . \tag{2.9}$$

Using the formula $\sum_{X=0}^{+\infty} q^X = (1-q)^{-1}$ for the infinite geometric row sum with $0 < q < 1$, the coefficient C_0 is calculated as

$$C_0 = \frac{1}{2} \left[p - 1 + \frac{1}{1 - \frac{p}{1-p}} \right]^{-1} = \frac{1-2p}{4p(1-p)} . \tag{2.10}$$

Taken together with Eq. (2.7), the stationary probability is completely determined as

$$P_X^s = \frac{1-2p}{4p(1-p)} \left(\frac{p}{1-p}\right)^X \qquad (1 \le X) \tag{2.11}$$

$$P_0^s = \frac{1-2p}{2(1-p)} , \tag{2.12}$$

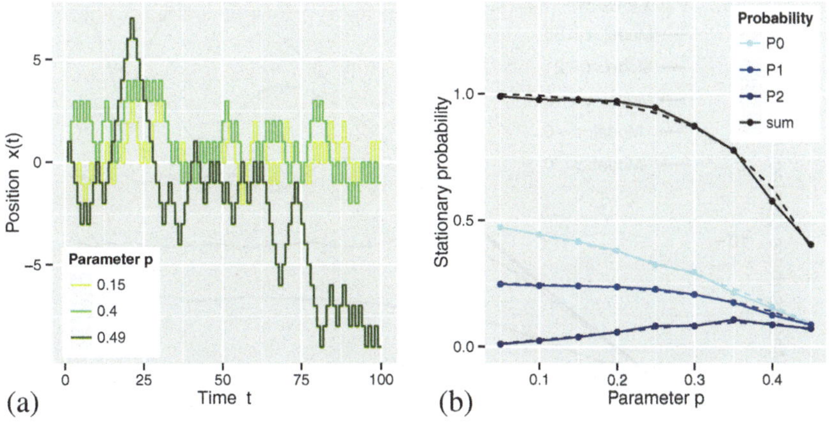

Fig. 2.7 Random walk of the model by Ohira and Milton without temporal delay. (**a**) Sample trajectories. (**b**) Comparison between numerical simulation of the stationary probability and the analytic solution

where $P_{-X} = P_X$ for all $1 \leq X$ due to the symmetry of the problem.

A comparison between simulated stationary probabilities and the analytic solution, Eqs. (2.11) and (2.12), illustrates the validity of the calculations (Fig. 2.7). Computing the sum of probabilities $P_0^s + P_1^s + P_2^s + P_{-1}^s + P_{-2}^s$ (black line in Fig. 2.7) we see that the stationary probability for $p \leq 0.3$ is dominated by the five states $\{0, \pm 1, \pm 2\}$. Furthermore, it is possible to calculate the variance [18] (see Problem 2.3) as

$$\sigma_0^2 = \frac{1}{2(1-2p)^2} \,. \tag{2.13}$$

In the next step of model development, Ohira and Milton introduce a *time delay* τ in the iteration rule, Eq. (2.3), i.e.,

$$Q(t) = \begin{cases} p & : X(t-\tau) > 0 \\ 1/2 & : X(t-\tau) = 0 \\ 1-p & : X(t-\tau) < 0 \end{cases} . \tag{2.14}$$

The position variance of this model can be solved by the same method, however, derivation of the corresponding analytic solutions involves tedious calculations. Numerical simulations of the two-point correlation function, computed for the estimator in Eq. (2.2), are shown in Fig. 2.8; results indicate that the model with temporal delay can explain the existence of two qualitatively different timescales.

Ohira and Milton [18] report that the dependence of the standard deviation of the fluctuation on the time delay τ and the parameter p can be approximated by

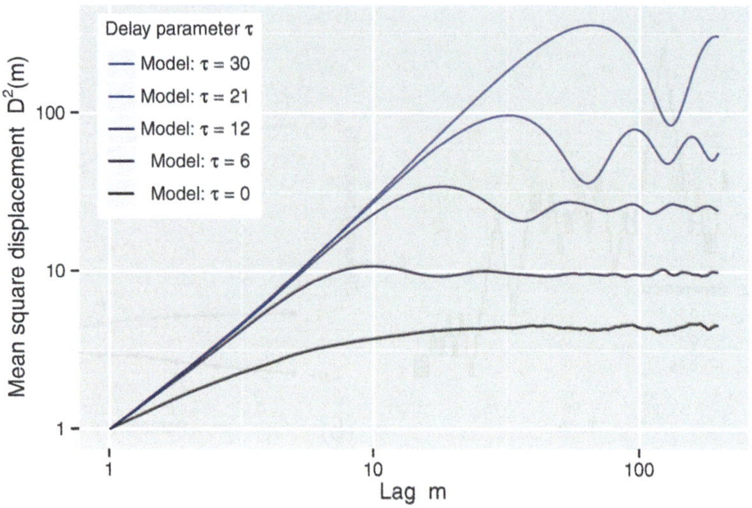

Fig. 2.8 *Delayed random walk* simulated by the model of Ohira and Milton [18]. The two-point correlation function, Eq. (2.2), of the model for $p = 0.25$ and different values for the *delay* τ approximates the qualitative transition between short and long timescales observed in experimental data

$$\sigma(\tau, p) \approx (0.59 - 1.18p)\tau + \frac{1}{\sqrt{2}(1-2p)} . \tag{2.15}$$

Therefore, measurement of σ can be used to estimate the parameters of the model.

Motivated by these results from Ohira and Milton's study [18], Mergenthaler and Engbert [16] proposed a random-walk model with nonlinear, time-delayed feedback for the control of fixational eye movements [16]. In this model, the different neural pathways of activations from *excitatory burst neurons* (EBN) and *tonic units* (TU) that control the eye muscles (see Fig. 2.2) are explicitly taken into account.

At time step i, it is assumed that the EBN activity w_i determines the eye's velocity, where an autoregressive term $(1-\gamma)w_i$ describing frictional damping and a term $-\lambda \tanh(\epsilon w_{i-\tau})$ that generates a nonlinear error correction with temporal delay τ are combined in an iteration rule (equation of motion), i.e.,

$$w_{i+1} = (1-\gamma)w_i + \xi_i - \lambda \tanh(\epsilon w_{i-\tau}) \tag{2.16}$$

$$x_{i+1} = x_i + w_{i+1} + \eta_i . \tag{2.17}$$

In the second equation, the position x_{i+1} is updated by the movement increments $w_{i+1} \cdot \Delta t$ (with $\Delta t = 1$) at each time step. Moreover, during fixational eye movements it is assumed that the activity of the TUs is not systematically related to eye position (because spatial excursion is small). Therefore, the activity of TUs is described by an additional noise term η_i in Eq. (2.17).

A comparison between experimental data and model simulations [16] indicates good agreement for the qualitative transition between persistence and antipersistence, where data analysis was carried out using *detrended fluctuation analysis* (DFA) [19], a more advanced method for computing the Hurst exponent on the long time scale. The model parameters were chosen as $\gamma = 0.25$, $\lambda = 0.15$, $\epsilon = 1.1$, and $\tau = 70$. The standard deviations of the noise terms were $\sigma = 0.075$ for ξ_i and $\rho = 0.35$ for η_i.

A limitation of time-delayed random-motion models is that the distinction between physiological drift and microsaccades (as suggested by inspection of experimental data) is neglected. In Sect. 2.5, we will discuss an integrated model of dynamical coupling between slow movements (physiological drift) and microsaccades.

2.4 Statistics of Microsaccades

For the analysis of correlations in fixational eye movements (Sect. 2.2), we simplified the problem by ignoring the distinction between two qualitatively different forms of motion, physiological drift and microsaccades. In this section, we start with the analysis of microsaccades and develop an algorithm for the detection of microsaccades in experimental data. The detection procedure is critical for the investigation of statistical properties of microsaccades and for the analysis of possible forms of interactions between both movement types.

Starting point for our detection algorithm is the observation that eye velocities are considerably higher during microsaccades than during physiological drift (Fig. 2.9). Therefore, in the first step of the detection procedure, we need to estimate eye velocities from eye-tracking data. Since experimental data from video-based eye-tracking systems are contaminated by observational noise, we apply a running-average filter of the form

$$\mathbf{v}_n = \frac{\mathbf{x}_{n+2} + \mathbf{x}_{n+1} - \mathbf{x}_{n-1} - \mathbf{x}_{n-2}}{6\Delta t} , \tag{2.18}$$

which computes velocity samples from five subsequent data samples.

In the second step, a threshold value for the detection of potential microsaccade epochs in 2D velocity space is computed. This is done separately for horizontal and vertical velocity components using estimators for the fluctuation ranges, σ_x and σ_y, defined as

$$\sigma_{x,y} = \sqrt{\langle \left(v_{x,y} - \tilde{v}_{x,y}\right)^2 \rangle} . \tag{2.19}$$

The operator $\langle . \rangle$ denotes the median estimator of the fluctuations, which is important to suppress a potential bias from high-velocity samples that occur during microsaccades. The detection thresholds $\eta_{x,y}$ are chosen as multiples of the fluctuation ranges (Fig. 2.9b, dashed blue line),

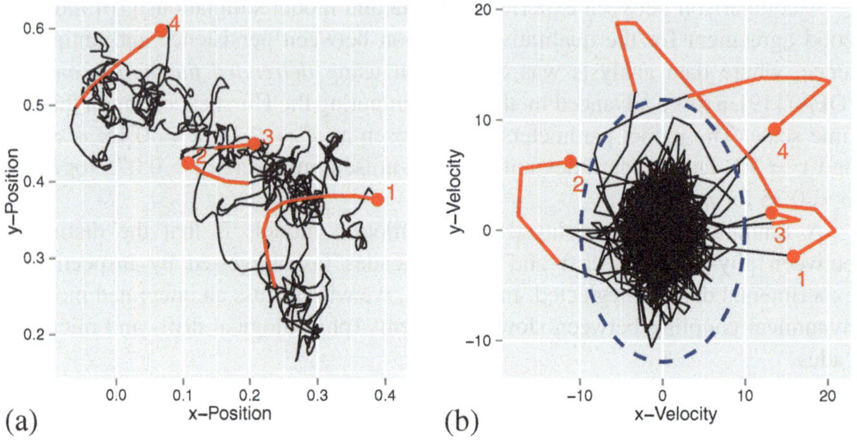

Fig. 2.9 Fixational eye movements and microsaccades (after: [10]). (**a**) Typical eye trajectory recorded during visual fixation with a duration of 2 s. Microsaccades (red color) are rapid, small-amplitude ballistic movements. Red numbers are related to endpoints of microsaccades. (**b**) A plot of the same trajectory in velocity space (y-velocity vs. x-velocity component) demonstrates that microsaccades are related to high-velocity epochs. The value of the detection threshold for microsaccades (dashed blue line) is chosen relative to the standard deviation of the eye's velocity

$$\eta_{x,y} = \lambda\sigma_{x,y}, \tag{2.20}$$

where λ is a free parameter for the detection algorithm that needs to be chosen appropriately.

In the third step, we can identify all data samples with velocities higher than the threshold, Eq. (2.20). For these data samples k, the test function $t(k)$ has the property

$$t(k) = \left(\frac{v_{k,x}}{\eta_x}\right)^2 + \left(\frac{v_{k,y}}{\eta_y}\right)^2 > 1\ . \tag{2.21}$$

Physiologically meaningful microsaccades are characterized by a minimum duration of about 5 ms. Depending on the sampling rate of the eye-tracking device, this minimum duration translates into different numbers of data points. For the datasets that we will use in the following, the criterion was chosen as a minimum of three subsequent data samples, for which the condition in Eq. (2.21) needs to be fulfilled, which is equivalent to a minimum duration of 6 ms.

In the fourth and final step, we consider only *binocular* events, i.e., microsaccades occurring in both eyes with a temporal overlap of at least one data sample. The binocularity is a consequence of the physiological assumption that microsaccades under central control by the nervous system. While the binocularity criterion is easy to check, it introduces a tremendous reduction of the noise in the detection procedure. Consider a sequence of data samples (a candidate sequence) recorded

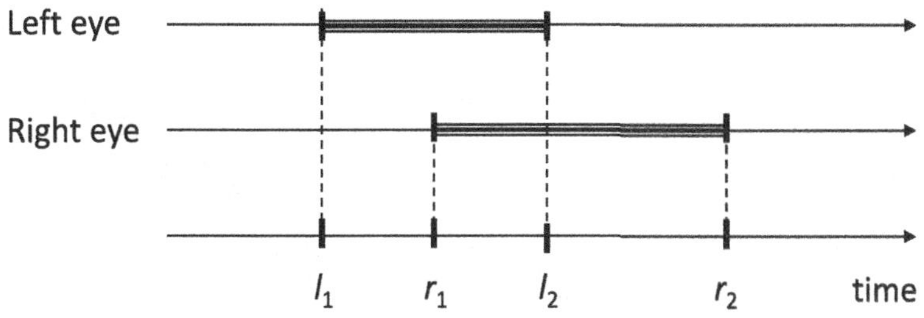

Fig. 2.10 Binocularity criterion. With onset times (r_1, l_1 for right and left eyes, resp.) and offsets (r_2, l_2), the overlap between microsaccade epochs from right and left eyes can be tested using Eq. (2.22)

from the right eye, where the eye's velocity in above threshold between time r_1 and time r_2. We can easily find overlap with the left eye beginning at time l_1 and ending at l_2 (Fig. 2.10) by the inequalities

$$r_2 > l_1 \qquad \text{and} \qquad r_1 < l_2 \,. \tag{2.22}$$

If these inequalities hold for two sequences in the left and right eyes, then there will be temporal overlap and the resulting sequence is chosen as a microsaccade.

A unifying property of saccades and microsaccades that plays an important role in the physiological literature is the so-called *main sequence* [22]. The main sequence is obtained by plotting the maximum velocity of the eye during a saccade (*peak velocity*) against the amplitude of the saccade, often in double-logarithmic scaling. Typically, a high correlation between the two kinematic parameters is observed (Fig. 2.11). The physiological interpretation of the main sequence is that microsaccades are a ballistic form of motion.

An important problem for the development of a theoretical model of fixational eye movements and microsaccades is related to possible interactions between slow movement components and microsaccades. In the absence of such a coupling, fixational eye movements could be generated as a simple superposition of two independent components, slow physiological drift and microsaccades. If the latter description is wrong, i.e., there are statistical interactions between drift and microsaccades, then a dynamical model for the generation of drift and microsaccades would be necessary. Moreover, it would be possible to predict microsaccades from modulations within the physiological drift movements.

Experimentally, it turned out that there is reduced movement activity immediately before a microsaccade [10]. To show this effect, the movement activity in different time intervals before a microsaccade is compared to random time intervals of the same lengths (i.e., without a systematic relation to microsaccade onset). An interesting quantitative measure for the eye's movement activity is purely geometrical and determines the number of boxes needed to cover the trajectory over a time window of 50 ms (Fig. 2.12a). For this *box-counting* procedure, the size of the boxes was chosen as the size of receptive field of retinal ganglion cells (linear

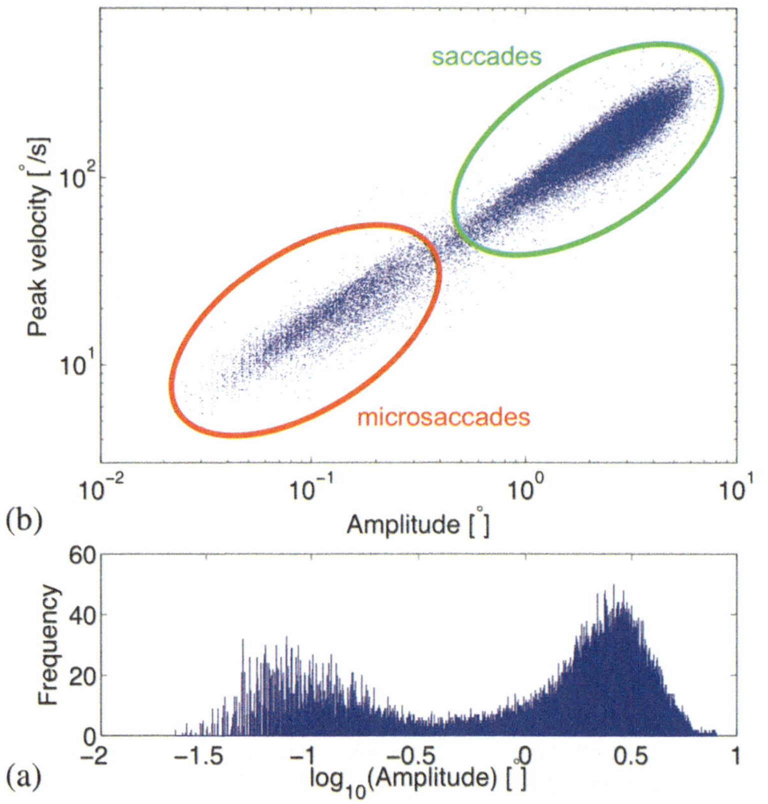

Fig. 2.11 Saccades and microsaccades. (**a**) The *main sequence*, a fixed relation between saccade amplitudes and peak velocities, extends to the microsaccade range. (**b**) Histogram of logarithmic saccade amplitudes from reading experiments

dimension $l \approx 0.03°$). The differences in box-counts before microsaccades and in randomly chosen intervals are plotted over a range from -500 ms to -100 ms prior to microsaccade onset. The difference is significantly negative around -200 ms before microsaccades (Fig. 2.12b), i.e., physiological drift is reduced before the next microsaccade. This effect is compatible with the interpretation that microsaccades represent an active process that prevents image fading by enhancing motion after epochs of reduced eye drift [6]. In the following, we will explore the corresponding hypothesis that microsaccades are triggered by low retinal image motion in a mathematical model.

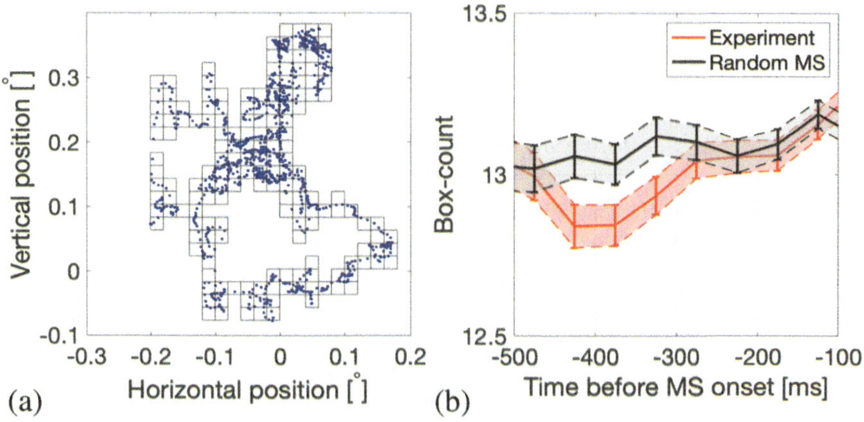

Fig. 2.12 Coupling of physiological drift and microsaccades [6]. (**a**) In the *box-counting* approach, the number of boxes needed to cover the trajectory over a certain time interval is determined. (**b**) Using a running window, the box count at different times before microsaccade onset is computed

2.5 An Integrative Model

An ideal starting point for the development of mathematical models for fixational eye movements is the observation of spatial and temporal correlations in high-precision eye-tracking. A major challenge for more advanced models is the integration of microsaccades and their statistical properties, since microsaccades might involve physiological systems other than those controlling slow movements. Thus, mathematical modeling might be used as a new tool to generate hypotheses on the control of microsaccades and their interaction with slower physiological drift movements.

The analysis and modeling of correlations in fixational eye movements, presented in Sects. 2.2 and 2.4, indicated persistent behavior on a short timescale, i.e., there is, at each point in time, the statistical tendency of the eyes to maintain their current movement direction. Such a behavior is observed in *self-avoiding random walks*, which are random motions with memory that produce autocorrelated behavior. In its strict mathematical form, a self-avoiding walk is terminated as soon the trajectory intersects itself. As a model for biological motion, we relax the self-termination principle and study statistically self-avoiding walks that show the tendency to stay away from recently visited locations.

An example for a possible model of a statistically self-avoiding walk was introduced by Freund and Grassberger in 1992 as a deterministic model of evolutionary change [12]. We will use their model as a starting point for the development of our mathematical model of fixational eye movements. The random motion is simulated on a square lattice of dimension $L \times L$ with periodic boundary conditions. Freund and Grassberger motivated their assumptions by considering a walk in a swamp.

A walker at current position (i, j) sinks as if he is standing in a swamp. As a consequence, the depth h_{ij} of the ground increases by a fixed amount per time unit in each iteration,

$$h_{ij} \mapsto h_{ij} + 1 \qquad \text{(Rule I: activation)} \tag{2.23}$$

at current position (i, j) of the lattice. The depth h_{ij} might be looked upon as a general variable describing some form of (neural) activation. As in a real swamp, activations at all other lattice positions (k, l) with $(k, l) \neq (i, j)$ decay according to the relaxation rule

$$h_{kl} \mapsto (1 - \epsilon) \cdot h_{kl} \, , \qquad \text{(Rule II: relaxation)} \tag{2.24}$$

where the relaxation rate ϵ is assumed to have a small value, i.e., the relaxation is much slower than the sinking of the walker. Equation (2.24) represents the special case of a linear decay of activation, the more general situation is discussed in the original work [12]. A visualization of rules I and II is presented in Fig. 2.13.

As a mechanism for the walker's movements, Freund and Grassberger introduced the iteration rule that, at each time step, the walker has to move to the one of its four neighboring lattice positions $\{(i - 1, j), (i, j - 1), (i + 1, j), (i, j + 1)\}$ with the minimum activation. Since it can happen that two or more activations are equal, we extend the rule by the stochastic principle that a random site among the set of neighboring sites with equal activation is chosen with equal probability.

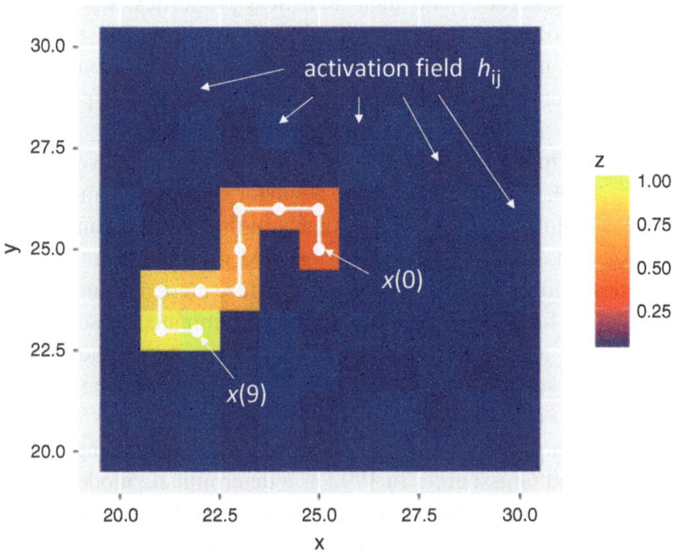

Fig. 2.13 Iteration rules of the self-avoiding walk [12]. Color indicates the activation at the lattice sites (i, j). Activation at $x(9)$ is increased by $+1$, all other sites relax

Following Freund and Grassberger [12], it is important to note that the resulting form of random motion is fully determined by a shortsighted rule (spatially and temporally local), i.e., there is no long-term strategy for the control of motion, and the decision in each movement step is completely determined by the value of the activation at the current site and the activations of the four neighbors. As a result of the iteration rules, the walker avoids to return to recently visited lattice positions and a self-organized distribution of activation is generated over the lattice.

Physiologically, the activation field $\{h_{ij}\}$ $(1 \leq i, j \leq L)$ represents neural activation of cells of the map motor in the corresponding layer of the superior colliculus controlling eye movements in the brainstem (see Sect. 2.1). What is needed in addition to the iteration rules of the self-avoiding walk, Eqs. (2.23) and (2.24), is a mechanism to keep the fixation close to an intended target position. Experimentally, it is observed that human participants are able to fixate an intended position within the foveal region ($\approx 2^\circ$ visual angle) for a duration of tens of seconds. Therefore, we add a (quadratic) potential [11] to the model, i.e.,

$$u(i, j) = \lambda L \left(\left(\frac{i - i_0}{i_0} \right)^2 + \left(\frac{j - j_0}{j_0} \right)^2 \right) , \tag{2.25}$$

where the coordinates (i_0, j_0) indicate the center of the lattice as the intended fixation position and the parameter λ determines the slope of the potential. As the revised iteration rule of the self-avoiding walk in the potential, the walker always moves to the position with the minimum of the sum of the self-generated activation field and the potential across the four neighboring sites. Thus, the new position (i', j') is given by

$$(i', j') = \arg \min_{(k,l) \in N(i,j)} \{h_{kl} + u(k, l)\} , \tag{2.26}$$

where $N(i, j) = \{(k, l) | k = i \pm 1 \text{ and } l = j \pm 1\}$. As a consequence, there are two driving forces in our random-walk model (Fig. 2.14).

First, the potential $u(i, j)$ drives the walker back to recently visited lattice positions to maintain fixation, which generates antipersistence. Second, the activation field h_{ij} produces the persistent part of the motion. The interaction of both driving forces must be studied in computer simulations. Our numerical simulations indicate that the transition from persistence on a short timescale to antipersistence on a longer timescale is qualitatively reproduced by the self-avoiding walk in the potential.

In the framework of an activation-based self-avoiding walk, it is possible to integrate microsaccades as an additional form of eye movements. In our model, the walker moves on the 2D lattice and visits positions with fluctuating values of activations, while positions with high activations are avoided on average. Using numerical simulations, it can be shown that the system self-organizes into a stationary state of the activation field. The corresponding stationary distribution will become more and more right-skewed with increasing values of the slope

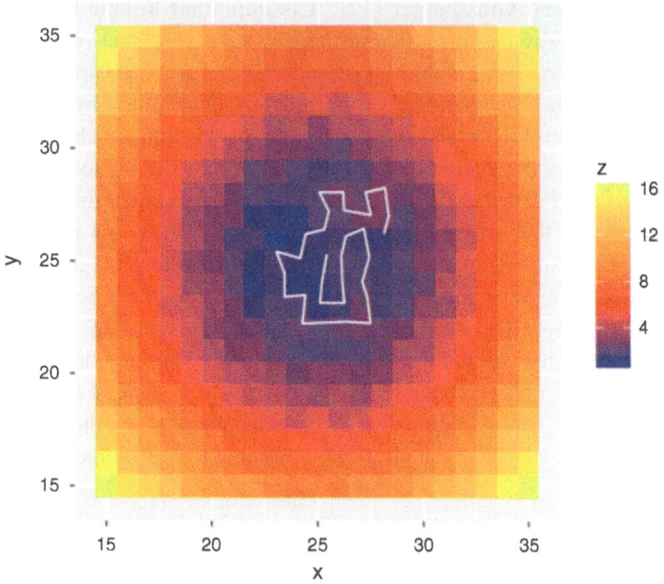

Fig. 2.14 Illustration of the simulation of the *self-avoiding walk* in a quadratic potential

parameter λ. Using the distribution of activations $\{h_{ij}\}$, a critical value of activation h_c can be defined, so that a jump to the global minimum of activation is generated whenever a lattice position with $h_{ij} > h_c$ is entered. This jump to the global minimum is interpreted as a microsaccade in the model (Fig. 2.15).

As an important property of fixational eye movements, we discussed that reduced fixational movements are observed immediately before microsaccades [6]. In our model, a microsaccade is more likely, if the local activation in the vicinity of the current lattice position is higher than on average. Because of the activation mechanism of the self-avoiding walk, Eq. (2.23), the walker will face high activations at its neighboring lattice positions, if the walker stayed within a limited area of the lattice for a while. Because of this built-in mechanism, the model should be able to reproduce the effect of reduced eye motion (retinal image slip) before microsaccades. Applying the box-counting procedure to simulated data, we were able to show that reduced retinal image slip is observed immediately before upcoming microsaccades [11]. Thus, the model not only integrates two very different forms of motion (slow movements and microsaccades) but also reproduces statistical relations between slow movements and microsaccades.

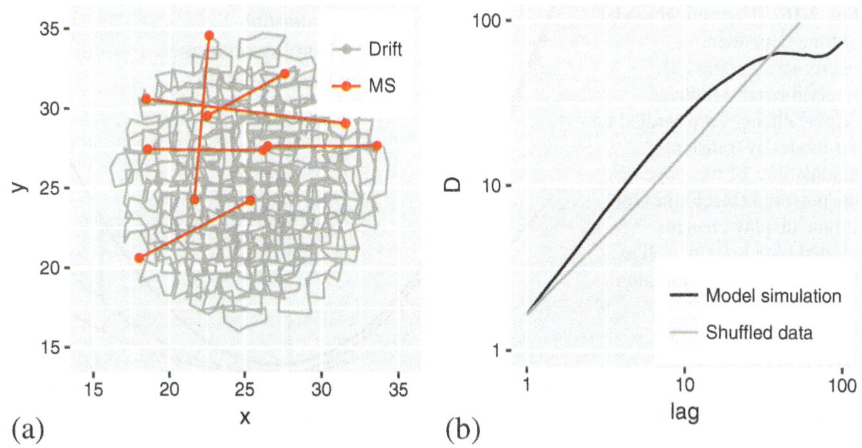

Fig. 2.15 Simulation of the *self-avoiding walk* in a quadratic potential. (**a**) Illustration of a simulated trajectory with microsaccades triggered by overcritical activations. (**b**) Using a plot of the mean square displacement $D^2(l)$ as a function of the time lag l, we observe that the model reproduces the transition from persistence to antipersistence found in experimental data

2.6 *Simulation of Time-Dependent Microsaccade Rates

In the last section of this chapter, we discuss how the integrated model of fixational eye movements can be used to study time-dependent microsaccade rates discussed in Sect. 1.4. It is observed in experiments that simple display changes as well as shifts in covert attention induced by the onset of visual stimuli produce characteristic modulations of the microsaccade rate (for an overview see [6]). Since it is the potential function $u(i, j)$ of our model that determines fixation position by allocation of attention to an intended target, a straightforward modeling assumption is that temporal variation of the microsaccade rate can be produced by the model via transient changes of the potential [7].

For the changes in microsaccade rates that are induced by display changes, we studied transient changes of the slope parameter λ of the movement potential, which could reproduce the time-dependent microsaccade rate effects from experiments. A transient change of the potential (Fig. 2.16, black line) by a reduction of the slope (blue line) was used as an unspecific response to a display change. As a result, the model generated the microsaccadic inhibition effect [7], since the transient widening of the potential induced a reduction of the average activation, so that overcritical activations were less frequent.

The subsequent short-term enhancement of the microsaccade rate could be obtained in the model, if a time-delayed reduction of the threshold h_c for microsaccade triggering was implemented in the model. For the estimation of the microsaccade rate, we applied the causal filter discussed in Sect. 1.4. Using parametric

Fig. 2.16 Theoretical model of time-dependent microsaccade rates [7]. Experimentally-induced display changes are studied in the model by transient modulations of the slope of the potential (black line) for simple display changes (dotted blue line) as well as for shifts of covert attention (dashed red line)

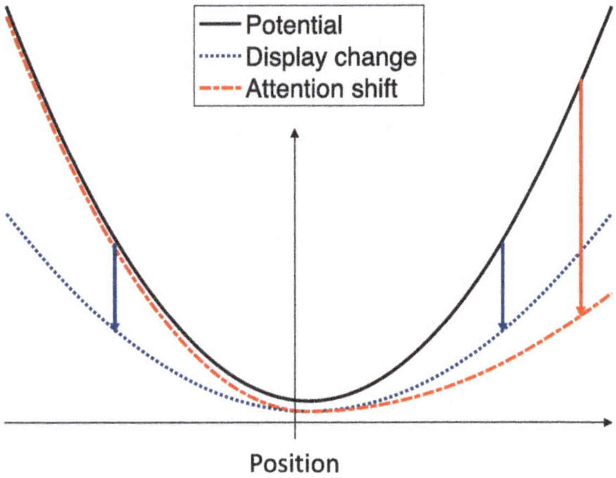

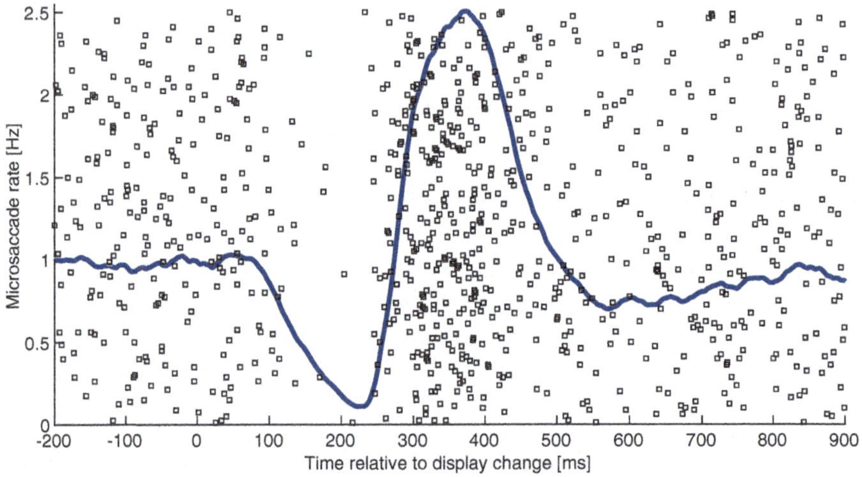

Fig. 2.17 Simulation of time-dependent microsaccade rates in the integrative model. Microsaccade onsets are indicated by the dots, plotted for single-trial simulations of the model in different rows; microsaccade rate (blue line) was estimated using a causal filter

changes in the model, we observed changes in microsaccade rates in the model that could reproduce effects observed in experimental data (Fig. 2.17).

Finally, it should be mentioned that the integrative model discussed so far can also be used to model effects of selective visual attention on microsaccade rates and orientations [7]. In this approach, asymmetrical changes of the movement potential (Fig. 2.16, red line) were studied to induce changes in the angular distribution of microsaccades. As a result, we observed the complicated time-dependent changes of microsaccade rates and orientations in response to display changes and attentional cues that were found in experiments [6]. In summary, time-dependent microsaccade rate effects in the model are qualitatively and quantitatively in good agreement with experimental data.

2.7 Exercises

Exercise 2.1 (Analysis of Fixational Eye Movements) (a) Use the estimator in Eq. (2.2) to plot the mean square displacement as a function of time lag for experimental data (stored in file `data.zip`). (b) Check the statistical validity of the result using computer-generated surrogate data that can be obtained by the following procedure: (1) Compute the movement increments. (2) Randomize the order of the increments. (3) Compute the cumulative sum of increments to generate a surrogate trajectory. Why is this an interesting null model of the experimental data?

Exercise 2.2 (Numerical Simulation of the Ohira and Milton Model) (a) Simulate the random walk of the Ohira-Milton model without time delay. Compute the stationary probabilities for the five states $\{0, \pm 1, \pm 2\}$. Compare the numerical result with the analytic solution, Eqs. (2.7) and (2.10), as in Fig. 2.7. (b) Simulate the model for different values of the time delay τ and compute the two-point correlation function using the estimator for $D^2(m)$ at *lag* m from Eq. (2.2) as in Fig. 2.8. *(Hint: Use the computer code for the estimator used in Problem 2.1.)*

Exercise 2.3 (Variance of the Ohira-Milton Model) Derive the analytical result, Eq. (2.13), for the variance of the Ohira-Milton model [18] for delay $\tau = 0$. (*Hint: Use the definition of the expectation value and the symmetry of the problem; note also that* $\sum_{x=1}^{\infty} x^2 q^x = q(1+q)/(1-q)^3$ *for* $|q| < 1$).

Exercise 2.4 (Microsaccade Detection from Experimental Data) Apply the detection algorithm for microsaccades to experimental data. Investigate the influence of the binocularity criterion, Eq. (2.22), on the resulting microsaccade rates.

Exercise 2.5 (Simulation of the Integrated Model of Fixational Eye Movements) (a) Investigate the transition from persistence to antipersistence in the model by Engbert et al. [11] for several sets of model parameters. (b) Investigate the influence of microsaccades on persistence and antipersistence of the model.

2.8 Further Reading

A number of review articles on research in fixational eye movements and microsaccades have been published [6, 15]. A central function of fixational eye movements has been proposed by Michele Rucci et al. [20] by assuming that fixational eye movements decorrelate the input from neighboring receptive fields. In an important theoretical work of visual functioning, Burak et al. [2] proposed a model using Bayesian estimation of dynamical image processing in the presence of fixational eye movements.

References

1. Brown, R. (1828). A brief account of microscopical observations made in the months of June, July and August 1827, on the particles contained in the pollen of plants; and on the general existence of active molecules in organic and inorganic bodies. *The Philosophical Magazine, 4*(21), 161–173. https://doi.org/10.1080/14786442808674769
2. Burak, Y., Rokni, U., Meister, M., & Sompolinsky, H. (2010). Bayesian model of dynamic image stabilization in the visual system. *Proceedings of the National Academy of Sciences, 107*(45), 19525–19530. https://doi.org/10.1073/pnas.1006076107
3. Collins, J. J., & De Luca, C. J. (1993). Open-loop and closed-loop control of posture: a random-walk analysis of center-of-pressure trajectories. *Experimental Brain Research, 95*(2), 308–318. https://doi.org/10.1007/BF00229788.
4. Coppola, D., & Purves, D. (1996). The extraordinarily rapid disappearance of entopic images. *Proceedings of the National Academy of Sciences, 93*(15), 8001–8004. https://doi.org/10.1073/pnas.93.15.8001
5. Einstein, A. (1905). Über die von der molekularkinetischen Theorie der Wärme geforderte Bewegung von in ruhenden Flüssigkeiten suspendierten Teilchen. *Annalen der Physik, 322*(8), 549–560. http://onlinelibrary.wiley.com/doi/10.1002/andp.19053220806/epdf
6. Engbert, R. (2006). Microsaccades: A microcosm for research on oculomotor control, attention, and visual perception. *Progress in Brain Research, 154*, 177–192. https://doi.org/10.1016/S0079-6123(06)54009-9
7. Engbert, R. (2012). Computational modeling of collicular integration of perceptual responses and attention in microsaccades. *The Journal of Neuroscience, 32*(23), 8035–8039. https://doi.org/10.1523/JNEUROSCI.0808-12.2012
8. Engbert, R., & Kliegl, R. (2003). Microsaccades uncover the orientation of covert attention. *Vision Research, 43*, 1035–1045. https://doi.org/10.1016/S0042-6989(03)00084-1
9. Engbert, R., & Kliegl, R. (2004). Microsaccades keep the eyes' balance during fixation. *Psychological Science, 15*(6), 431. https://doi.org/10.1111/j.0956-7976.2004.00697.x
10. Engbert, R., & Mergenthaler, K. (2006). Microsaccades are triggered by low retinal image slip. *Proceedings of the National Academy of Sciences, 103*(18), 7192–7197. https://doi.org/10.1073/pnas.0509557103
11. Engbert, R., Mergenthaler, K., Sinn, P., & Pikovsky, A. (2011). An integrated model of fixational eye movements and microsaccades. *Proceedings of the National Academy of Sciences, 108*(39), E765–E770. https://doi.org/10.1073/pnas.1102730108
12. Freund, H., & Grassberger, P. (1992). The red queen's walk. *Physica A: Statistical Mechanics and its Applications, 190*(3), 218–237. https://doi.org/10.1016/0378-4371(92)90033-M
13. Klafter, J., & Sokolov, I. M. (2011). *First steps in random walks: from tools to applications*. Oxford: Oxford University Press. https://global.oup.com/academic/product/first-steps-in-random-walks-9780199234868
14. Lettvin, J. Y., Maturana, H. R., McCulloch, W. S., & Pitts, W. H. (1959). What the frog's eye tells the frog's brain. *Proceedings of the IRE, 47*(11), 1940–1951. https://doi.org/10.1109/JRPROC.1959.287207
15. Martinez-Conde, S., Macknik, S. L., & Hubel, D. H. (2004). The role of fixational eye movements in visual perception. *Nature Reviews Neuroscience, 5*(3), 229–240. https://doi.org/10.1038/nrn1348
16. Mergenthaler, K., & Engbert, R. (2007). Modeling the control of fixational eye movements with neurophysiological delays. *Physical Review Letters, 98*(13), 138104. https://doi.org/10.1103/PhysRevLett.98.138104
17. Metzler, R., & Klafter, J. (2000). The random walk's guide to anomalous diffusion: a fractional dynamics approach. *Physics Reports, 339*(1), 1–77. https://doi.org/10.1016/S0370-1573(00)00070-3
18. Ohira, T., & Milton, J. G. (1995). Delayed random walks. *Physical Review E, 52*(3), 3277. https://doi.org/10.1103/PhysRevE.52.3277

19. Peng, C.-K., Buldyrev, S. V., Havlin, S., Simons, M., Stanley, H. E., & Goldberger, A. L. (1994). Mosaic organization of DNA nucleotides. *Physical Review E, 49*(2), 1685. https://doi.org/10.1103/PhysRevE.49.1685
20. Rucci, M., & Casile, A. (2004). Decorrelation of neural activity during fixational instability: Possible implications for the refinement of v1 receptive fields. *Visual Neuroscience, 21*(5), 725–738. https://doi.org/10.1017/S0952523804215073
21. Sparks, D. L. (2002). The brainstem control of saccadic eye movements. *Nature Reviews Neuroscience, 3*(12), 952–964. https://doi.org/10.1038/nrn986
22. Zuber, B. L., Stark, L., & Cook, G. (1965). Microsaccades and the velocity–amplitude relationship for saccadic eye movements. *Science, 150*, 1459–1460. https://doi.org/10.1126/science.150.3702.1459

Chapter 3
Information Accumulation in Simple Decisions

Human choice behavior is investigated in the field of decision making. Here we discuss two-alternative forced choice as an example for a simple decision task. Experimentally, we are faced with two qualitatively different measures of human performance. First, we observe certain *response probabilities*, i.e., the probabilities of responses (e.g., "right" versus "wrong" or "signal" versus "noise"). Second, each response is measured after a *reaction time*, which is the time difference between the time of the response (e.g., keypress or saccadic eye movement) and stimulus onset. In many problems in experimental psychology, response probabilities or reaction times are considered as a single measure of performance. However, for a full theoretical analysis of human behavior in simple decision tasks, a joint model of both response probabilities and reaction times needs to be developed, which is the topic of this chapter. We follow the discussion given in the textbook by Busemeyer and Diederich [3], in particular, in Sects. 3.1 and 3.2.

An important example for the simple decision task studied in the following is the signal-detection paradigm. A participant is asked to judge the presence or absence of a weak signal that is overlayed by noise. In corresponding real-world decision tasks, a two-alternative response must be carried out—often there could be serious consequences. You might think of the distinction between an armed or weaponless person approaching a police checkpoint under dim illumination. Table 3.1 introduces the nomenclature of the resulting four combinations of two stimuli and two responses of a participant in an experiment.

A very successful class of theoretical models for simple decisions is based on the assumption that all relevant information is acquired sequentially. The sequential uptake of information is accumulated until a decision criterion is reached which triggers a motor response. Accumulator models are process models that can be applied to a variety of situations. For example, older participants show prolonged response times to a stimulus. One hypothesis is that the rate of information accumulation is reduced in older subjects. Second, the threshold representing the decision criterion could be increased, or, third, the motor response could be slowed.

R. Engbert, *Dynamical Models In Neurocognitive Psychology*,
https://doi.org/10.1007/978-3-030-67299-7_3

Table 3.1 Nomenclature of the four combinations of presented stimuli and responses of the participant in a signal-detection task

Participant's response	Signal condition	Noise condition
Signal	Correct (*hit*)	False (*false alarm*)
Noise	False (*miss*)	Correct (*correct rejection*)

Interestingly, each of the modeling assumptions will generate specific patterns of results related to response probabilities and reaction times.

In the case of reactions carried out via saccadic eye movements, neurophysiological structures for the control of activation in saccade generation, e.g., the superior colliculus (see Chap. 2, Fig. 2.3 on page 20), show build-up of activation that precedes a saccade. At the end of this chapter, we will relate the mathematical models to more recent results on the neurophysiology underlying simple decisions.

3.1 Models of Information Accumulation

In an accumulator model, the decision process starts with an initial value $X(0) = X(t = 0)$ at time $t = 0$ (see [3]). Over time, accumulation of information increases by sensory processing of a stimulus. In time step $h > 0$, the amount of evidence has changed by the increment $V(h)$ to $X(h) = X(0) + V(h)$. If this change is positive, $V(h) > 0$, then there is a tendency to the "signal" response. A negative change, $V(h) < 0$, indicates a tendency toward the "noise" response. After n steps at time $t = n \cdot h$, the accumulator is in state

$$X(t) = X(t - h) + V(t) \tag{3.1}$$

$$= X(0) + \sum_{j=1}^{n} V(j \cdot h) = X(0) + S(t) \, , \tag{3.2}$$

where the accumulated evidence at time t has reached the value $S(t)$.

In this model, variation of the initial value $X(0)$ represents *a priori* knowledge that favors one of the two-alternative decisions. Additionally, experimental research on neural dynamics in decision making demonstrated that influences from previous trials will modulate neural activations in the current trial [4]. However, such complications that can, in principle, be included in the model are neglected in this chapter.

In the accumulator model, the decision criterion consists of an upper threshold, $+\theta_S > 0$, for a signal response and a lower threshold, $-\theta_N < 0$, for a noise response. The response is initiated as soon as one of the two thresholds is reached for the first time. At this point, the accumulation stops. Therefore, the states $+\theta_S$ and $-\theta_N$ are denoted as *absorbing states*. The corresponding *stopping rule* is given by

$$-\theta_N < X(t) < +\theta_S : \text{Continue accumulation} \tag{3.3}$$

$$X(t) \geq +\theta_S : \text{Stop; signal response} \tag{3.4}$$

$$X(t) \leq -\theta_N : \text{Stop; noise response.} \tag{3.5}$$

Within this framework, we can study both response probabilities (i.e., probabilities to reach one of the two thresholds) and decision times (i.e., time to reach one of the two thresholds). In general, the model produces dependencies between the two types of variables, response times and response probabilities, which are typically measured in experiments.

In general, the distances of the two thresholds, $+\theta_S$ and $-\theta_N$, from the initial value $X(0)$ are different or asymmetrical, $\theta_S \neq \theta_N$. The following analysis demonstrates that this asymmetry can be replaced by symmetrical thresholds with a systematically shifted initial value (*payoff-bias*), $X(0) \neq 0$, and a relation between speed and accuracy of the response (*speed-accuracy trade-off*). These new parameters are defined as

$$\text{(payoff-bias) } b = \frac{1}{2}(\theta_N - \theta_S) \tag{3.6}$$

$$\text{(speed-accuracy trade-off) } \theta = \frac{1}{2}(\theta_N + \theta_S). \tag{3.7}$$

Using these definitions, Eqs. (3.3)–(3.5) of the accumulator model can be replaced by a mathematically equivalent formulation, i.e.,

$$-\theta < X(t) + b < +\theta : \text{Continue accumulation} \tag{3.8}$$

$$X(t) + b \geq +\theta : \text{Stop; signal response} \tag{3.9}$$

$$X(t) + b \leq -\theta : \text{Stop; noise response.} \tag{3.10}$$

Taken together with Eq. (3.2), we can write the accumulator model as

$$-\theta < \beta + S(t) < +\theta : \text{Continue accumulation} \tag{3.11}$$

$$\beta + S(t) \geq +\theta : \text{Stop; signal response} \tag{3.12}$$

$$\beta + S(t) \leq -\theta : \text{Stop; noise response,} \tag{3.13}$$

where the new parameter $\beta = b + X(0)$ was introduced (*response bias*).

The accumulated evidence $V(t)$ in time step t will tend to a signal or noise response and is mathematically represented as a Gaussian distributed stochastic process with properties

$$\mu = E[V(t)] \qquad \text{and} \tag{3.14}$$

$$\sigma^2 = E[(V(t) - \mu)^2]\,, \tag{3.15}$$

where parameter $d = \mu/\sigma$ is termed the *discriminability* of the process. In signal trials, we assume that the discriminability is positive, $d > 0$, while, for noise trials, the discriminability is negative, $d < 0$. The signal strength or intensity is given by $|d|$. To simplify the analysis, we assume that the variance of the evidence is normalized, $\sigma^2 = 1$.

In an important extension of the basic accumulator model [7], Ratcliff assumed that the discriminability varies stochastically from trial to trial with variance η^2. Such statistical variations in discriminability might be caused, for example, by fluctuations of a subject's attention. For the limit of infinitesimally small time steps, $h \to 0$, we obtain the diffusion model as a continuous-time stochastic model for information accumulation [7].

In addition to the information accumulation, models of simple decisions need to be completed by motor-level components to generate realistic response time distributions. In the example of saccadic decision, the motor-related activity is generated by neurons in the superior colliculus that project along the oculomotor pathway to the eye muscles (Fig. 2.2 on p. 19). Because of this motor component in reaction times, experimental measures do not represent the decision process itself—a fact that must be addressed in a full mathematical model, of course. The experimentally observed reaction time C_{RT} is given as the sum of the decision time D_T and the motor delay M_T, i.e.,

$$C_{RT} = D_T + M_T \,. \tag{3.16}$$

In human motor control, a basic finding is that the statistical properties of the motor delay M_T depend on the complexity of the task [9]. As a result, the variance of the reaction-time distribution can be dominated either by the variance of the decision time D_T (in the case of difficult signal-detection tasks in combination with simple motor responses) or by the variance of the motor delay M_T (in the case of simple signal-detection tasks with complicated motor responses).

In summary, we identified three different parameters in the basic dynamical model for signal detection: (1) The *response bias* β represents the *a priori* information of the participant. (2) The discriminability d represents the signal strength. It is positive in signal trials and negative in noise trials. (3) The distance between the two thresholds $\pm\theta$ controls the *speed-accuracy trade-off*. Finally, we must take into account that statistical fluctuations exist in both the discriminability and the motor response.

The class of stochastic models discussed so far is often denoted as *drift-diffusion models* or *information-accumulation models* [10]. For many of the model variants, there exist analytical solutions of the response probabilities and response time distributions that can be compared to experimental data. In the next section, we will study the mathematical solutions of the discrete random-walk model with constant time steps as a representative example for a decision model.

3.2 Elementary Solutions of the Random-Walk Model

In the discrete random-walk model for simple decisions, information accumulation is described by a stochastic process with constant step size in time ($h = 1$) and discrete states. For this model, we assume that the decision criterion is at $\pm\theta$ with $\theta = K$, i.e., there are $N = 2K + 1$ discrete states. To simplify the analytical calculations, we shift all states of the state space by $+K$, so that the state space consists of $N = 2K + 1$ elements ranging from 0 to $+2K$. The states are connected via one-step transitions ± 1, where the transition probabilities are p for $n \mapsto n + 1$ and $q = 1 - p$ for $n \mapsto n - 1$ (Fig. 3.1). Therefore, sampling of information with accumulation toward a signal response is performed with probability $0.5 < p < 1$, while the accumulator tends to a noise response with probability $0 < p < 0.5$. The discriminability of the accumulation process is related to the value of p. Since the random walk stops, if one of the two decision criteria 0 or $2K$ is reached, these states are called *absorbing* states. Before we compute the mathematical solutions of the model, we discuss some nomenclature of the underlying class of stochastic processes.

A sequence $\{X_n\}_{n\geq 0}$ of random variables with values taken from the set E is called a *temporally discrete stochastic process* with *state space E*. In the following, we assume that the set E is countable with elements $i, j, k, \ldots$. Let us assume that $\{X_n\}_{n\geq 0}$ is a temporally discrete stochastic process over state space E. If for all $n \geq 0$ and all states $i_0, i_1, \ldots, i_{n-1}, i, j$ it holds that

$$P(X_{n+1} = j|X_n = i, X_{n-1} = i_{n-1}, \ldots, X_0 = i_0) = P(X_{n+1} = j|X_n = i), \tag{3.17}$$

then the stochastic process is called a *Markov chain* [2]. If the right-hand side of Eq. (3.17) does not depend on n, then the process is a *homogeneous Markov chain*. The matrix $\mathbf{P} = \{p_{ij}\}_{i,j\in E}$ with

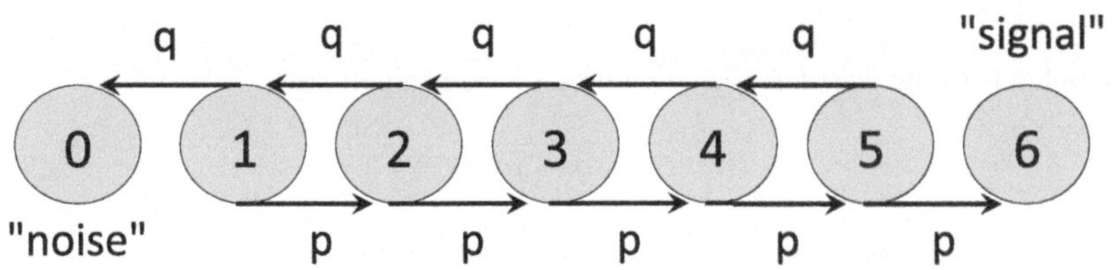

Fig. 3.1 Illustration of the discrete random-walk model. Diagram of the possible transitions for the model with $K = 3$, i.e., the state space is $E = \{0, 1, 2, .., 2K\}$. With transition probability p the step $n \mapsto n + 1$ is carried out, while the transition $n \mapsto n - 1$ is performed with probability $q = 1 - p$. The states $\{0; 2K\}$ are absorbing; these states represent the stopping rule, such that information accumulation terminates as soon as one of the states is reached. Simulation examples are shown in Fig. 3.2

$$p_{ij} = P(X_{n+1} = j | X_n = i) \tag{3.18}$$

is termed the *transition matrix* of the homogeneous Markov chain.

Now we consider the above random-walk model for simple decision, which is the homogeneous Markov chain over the state space $E = \{0, 1, 2, \ldots, 2K\}$, illustrated in Fig. 3.1. Here the decision criteria are represented by the absorbing state 0 for a noise response and by the absorbing $2K$ for the signal response. The decision time C_{RT} of the random walk is the first time step n, where one of the absorbing states $X_n = 0$ (noise) or $X_n = 2K$ (signal) is reached. We define the probability for the signal response as the conditional probability

$$c_i = P(X_T = 2K | X_0 = i) \qquad \text{for all} \qquad i \in [0, 2K] \, , \tag{3.19}$$

if sampling of information started at the initial state $X_0 = i$.

For the calculation of the probability to generate a signal response, Eq. (3.19), we will carry out a *first-step analysis* [2]. When the walker is at position i, it can move with probability p to $i + 1$ (from where the probability will be c_{i+1}) or with probability q to $i - 1$ (from where the probability will be c_{i-1}). Therefore, the probability c_i for a signal response from initial value i is related to the probabilities c_{i+1} and c_{i-1} via the equation

$$c_i = pc_{i+1} + qc_{i-1} \qquad \text{for} \qquad i = 1, 2, 3, \ldots, 2K - 1 \tag{3.20}$$

with boundary conditions

$$c_0 = 0 \qquad \text{and} \qquad c_{2K} = 1 \, . \tag{3.21}$$

To solve the system of difference equations of second order in Eq. (3.20), we choose the ansatz $c_i = x^i$ and try to determine the free parameter x. Substitution of the ansatz into Eq. (3.20) gives

$$x^i = qx^{i-1} + px^{i+1} \, , \tag{3.22}$$

which leads to the *characteristic equation* of the problem as

$$x = q + px^2 \, . \tag{3.23}$$

This equation can be solved and we obtain the two solutions

$$x = \frac{1 \pm \sqrt{1 - 4pq}}{2p} \, , \qquad \text{where} \qquad 1 - 4pq = (1 - 2p)^2 \, , \tag{3.24}$$

that can be simplified to the two solutions

$$x_1 = 1 \qquad \text{and} \qquad x_2 = \frac{q}{p} = \frac{1-p}{p}, \tag{3.25}$$

when $p \neq q$ (or a double root $x = 1$, if $p = q = 1/2$). The general solution of the system is obtained as a linear combination of the solutions, Eq. (3.25), which can be written as $c_i = a + b(q/p)^i$. Using the boundary conditions we calculate the free parameters a and b. From $c_0 = 0$, we obtain $a + b = 0$ or $b = -a$. Using $c_{2K} = 1$ it follows that $a = [1 - (q/p)^{2K}]^{-1}$. Having determined the free parameters, we are now able to write the solution for the marginal probability of the signal response from initial value i in closed form as

$$c_i = \frac{1 - (q/p)^i}{1 - (q/p)^{2K}}. \tag{3.26}$$

Using numerical simulations (Fig. 3.2), we can check the validity of the analytical result in Eq. (3.26). Beyond the formula for the marginal probability, we can also calculate the decision-time distribution [3]. Both problems are investigated in the exercises at the end of this chapter.

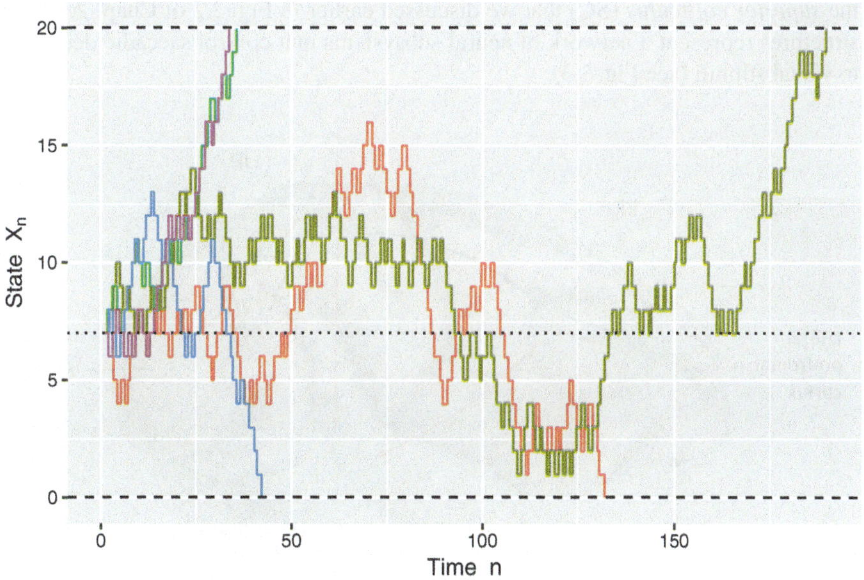

Fig. 3.2 Simulation of the discrete random-walk model. Ten sample trajectories are shown in different colors. Parameters were chosen as $K = 10$, $p = 0.55$ (or $q = 0.45$), and $i = 7$ (or $X(0) = -3$)

3.3 Neurophysiology of Decision Making

The idea that, in simple decisions, evidence is accumulated sequentially from elementary processes until a decision criterion is reached, was developed in cognitive and mathematical psychology [5, 6]. Today, a rich literature on experiments and mathematical models exists. When first models were developed, neural foundations were more or less unknown. This situation has dramatically changed recently, since substantial progress has been made on the study of neural activity using single-cell recording in the awake monkey. As a consequence, we observe the fusion of two areas of decision making, neuroscience and mathematical psychology [10], that were perceived as separate fields traditionally.

In neurobiological research on simple decision making in two-alternative choice tasks, experimental paradigms are typically similar to the paradigms developed in experimental psychology. For example, an animal is trained to respond to a motion stimulus presented on a computer display by a saccadic eye movement. Using single-cell recording from several cortical and subcortical areas, it could be shown that the activity of these neurons correlates with the ongoing decision process. Related neural areas are the *frontal eye fields* (FEF), the *lateral intraparietal* (LIP) area of the extrastriate cortex, the motion-sensitive *middle temporal* (MT) area, and the *superior colliculus* (SC) that we discussed earlier in Fig. 2.2 of Chap. 2. These structures represent a network of neural subsystems that control saccadic decisions to visual stimuli (see Fig. 3.3).

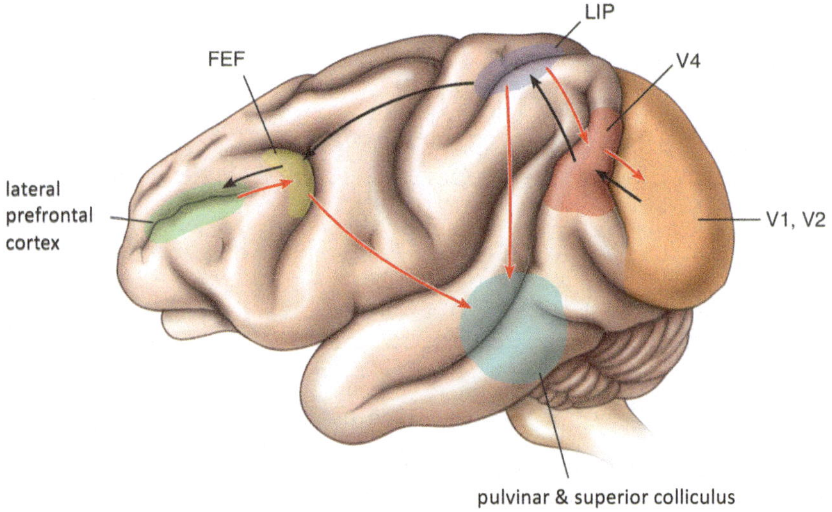

Fig. 3.3 Network of neural structures for the control of eye movements. Signal from the frontal eye fields (FEF) and the lateral intraparietal cortex (LIP) project to the superior colliculus (SC) as the top-level brainstem structure for saccade control. Each of the areas provides neurons with highly specific properties for the control of saccades (from: [1])

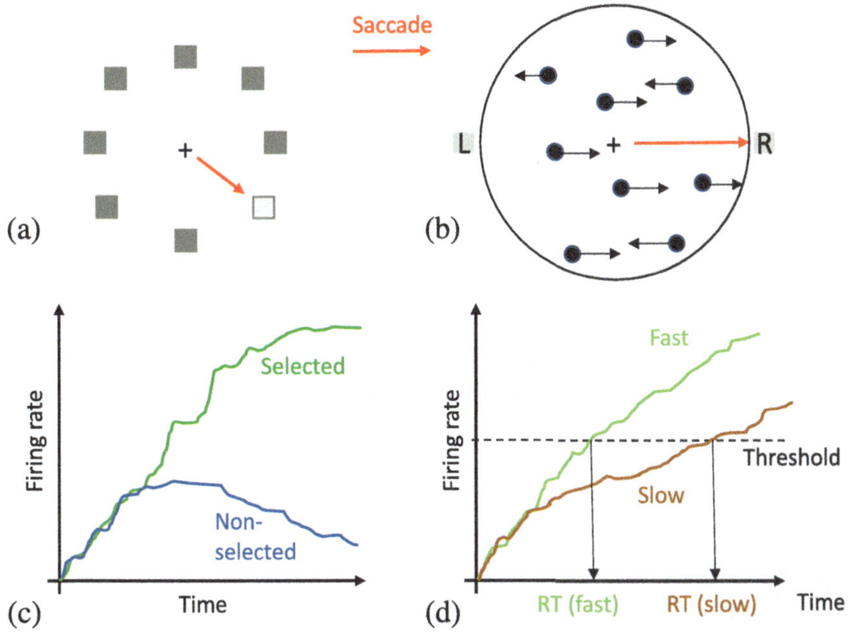

Fig. 3.4 Activity of single cells in animal experiments. (**a**) In the *oddball* paradigm monkeys are trained to performed a saccade to the deviant stimulus. (**b**) In the *coherent motion* paradigm, the animal is required to generate a saccade in the direction of the coherent motion stimulus. (**c**) Time-course of the firing rate for a typical neuron in one of the decision-related areas (see text). (**d**) The rise-rate of the firing rate predicts the saccadic reaction time of the animal on the level of single trials (after: [10])

In the example of the *oddball* paradigm, a monkey is trained to respond with a saccadic eye movement to a deviant stimulus, which differs from a series of distractors presented before (Fig. 3.4a). The critical stimulus (target) is different from the distractors with respect to one simple property (e.g., color, form, or contrast). Single-cell recording was carried out from neurons of the fontal eye fields (FEF). A distractor or target stimulus could be positioned inside or outside the receptive field of the neuron that was recorded from.

Data demonstrated that the presentation of an arbitrary stimulus (distractor or target) initially evoked increasing firing rates. After some time, however, the cell's activity developed specific differences across conditions: The activity of a neuron with a receptive field containing the target showed steadily increasing or constant activity, while the firing rate of a neuron with a distractor in its receptive field decreased after the initial rising phase (Fig. 3.4c). Qualitatively similar results were obtained for motion stimuli in area MT or in LIP. Such activation patterns were also found in *build-up* neurons in the SC.

In a motion discrimination task (*coherent motion paradigm*), the animal has to detect coherently moving dots among a number of distractors that move in random directions. After detection, a saccadic response must be performed in the direction of the coherent motion display (Fig. 3.4b). Experimental results [10] indicate that the activity of single cells in several areas is a neural correlate of the ongoing decision process. First, the rise of activity is independent of the question whether a saccade is made or not (or whether the correct target was hit). Therefore, the activity is not merely an indicator of the upcoming motor response, but is related to the selection process between alternative responses. Second, the rise-rate of the firing rate predicts reaction time (Fig. 3.4d): a steeper increase is correlated with shorter reaction times. Third, the reaction time can be predicted from the time it takes to reach the threshold. Fourth, it could be shown that the rise-rate of the firing rate is modulated by the difficulty of the task. Finally, the rise-rate depends on the *a priori* information that is available about the task.

In summary, experiments demonstrate that central assumptions made in mathematical models of sequential information accumulation are in good agreement with results from modern neurobiological research [10].

3.4 Exercises

Exercise 3.1 (Numerical Simulation of Response Probabilities) Check the analytical calculations on the conditional probability for generating a signal response, Eq. (3.26), in the simple random-walk model via numerical simulations using different values of the model parameters.

Exercise 3.2 (Numerical Simulation of the Response Time Distribution) In the textbook by Busemeyer and Diederich [3], the cumulative probability $P_i(N)$ for a signal response after N iterations is derived as

$$P_i(N) = \frac{2^{N+1}}{2K}(pq)^{\frac{N+1}{2}}\left(\frac{p}{q}\right)^{\frac{2K-i}{2}}\sum_{j=1}^{m}\cos^N\left(\frac{\pi j}{2K}\right)\sin\frac{\pi j(2K-i)}{2K}\sin\frac{\pi j}{2K},$$

where m is the number of transient states, $m = 2K - 1$, and i is the initial value, $i \in \{0, 1, \ldots, 2K\}$ with $i = X(0) + K$. Show the validity of the equation for $P_i(N)$. Choose $p = 0.6$ and $i = K$. Set $K = 20$ and simulate 10,000 trials. Plot the resulting simulated probability density and the analytical result into the same coordinate system for comparison (Fig. 3.5).

3.5 Further Reading

For the mathematical background on Markov chains we refer to the book by Brémaud [2]. The analysis of the discrete random-walk model is discussed in detail

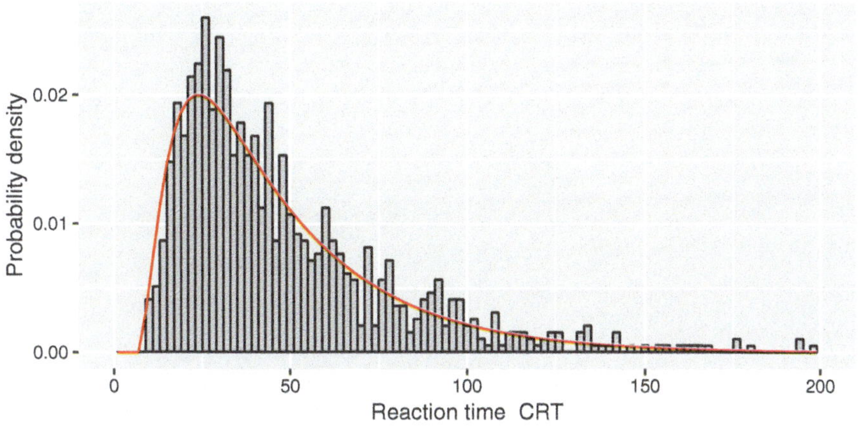

Fig. 3.5 Distribution of reaction times. The analytical solution (red line) for the reaction time distributions of the simple random-walk model in comparison to simulated data (histogram)

in the textbook by Busemeyer and Diederich [3], which also addresses the transition to the diffusion model [7]. An introduction to the diffusion model can be found in a review article by Ratcliff and McKoon [8].

References

1. Bear, M. F., Connors, B. W., & Paradiso, M. A. (2018). *Neurowissenschaften: Ein grundlegendes Lehrbuch für Biologie, Medizin und Psychologie*. New York: Springer-Verlag. https://link.springer.com/book/10.1007/978-3-662-57263-4
2. Brémaud, P. (1999). *Markov chains: Gibbs fields, Monte Carlo simulation, and queues* (Vol. 31). New York: Springer. http://www.springer.com/de/book/9780387985091#otherversion=9781441931313
3. Busemeyer, J. R., & Diederich, A. (2010). *Cognitive modeling*. Thousand Oaks: Sage. https://uk.sagepub.com/en-gb/eur/cognitive-modeling/Book226030
4. Fecteau, J. H., & Munoz, D. P. (2003). Exploring the consequences of the previous trial. *Nature Reviews Neuroscience, 4*(6), 435–443. https://doi.org/10.1038/nrn1114
5. Luce, R. D. (1959). *Individual choice behavior: A theoretical analysis*. Mineola, NY: Dover Publications.
6. Luce, R. D. (1986). *Response times: Their role in inferring elementary mental organization. Oxford psychology series* (Vol. 8). Oxford: Oxford University Press. https://global.oup.com/academic/product/response-times-9780195036428
7. Ratcliff, R. (1978). A theory of memory retrieval. *Psychological Review, 85*(2), 59–108. https://doi.org/10.1037/0033-295X.85.2.59
8. Ratcliff, R., & McKoon, G. (2008). The diffusion decision model: Theory and data for two-choice decision tasks. *Neural Computation, 20*(4), 873–922. https://doi.org/10.1162/neco.2008.12-06-420
9. Rosenbaum, D. A. (2009). *Human motor control*. Cambridge: Academic. http://psycnet.apa.org/psycinfo/2010-06119-000
10. Smith, P. L., & Ratcliff, R. (2004). Psychology and neurobiology of simple decisions. *Trends in Neurosciences, 27*(3), 161–168. https://doi.org/10.1016/j.tins.2004.01.006

Chapter 4
Sensorimotor Integration

A fundamental problem in neuroscience is the generation of adaptive behavior based on the integration of sensory signals and motor commands. In *sensorimotor integration* we investigate cognitive processing of sensory signals for the control of action [18]. In this case, sensory perception is the basis for the control of action. However, in Chap. 2, we learned that motor systems contribute to the acquisition of new sensory input, a phenomenon called *active perception*. Thus, sensory and motor systems are both fundamental neuro-cognitive systems that can provide a basis for each other, depending on task requirements.

4.1 The Crab Robin

The neurophysiological foundations of sensorimotor integration were discussed by Churchland in 1986 using a paradigmatic example [3]. Churchland introduced the artificial crab Robin as a (robotic) model organism that needs to solve sensorimotor integration implicitly. The crab is equipped with (1) two eyes that can be rotated to fixate a target object and with (2) a manipulator arm (Fig. 4.1). To perform goal-directed behavior using the manipulator arm, the crab needs to establish a functional relationship between the rotation angles of both eyes and the angles of the elements of the manipulator arm (see Exercise 4.1).

In this simple example, a two-dimensional (2D) sensory state space exists which represents the position of an arbitrary object that is focused by the eyes using the rotation angles (α, β). The crab must learn to generate an arm movement to all objects that are located within the reachable part of the environment. The manipulator arm is characterized by a 2D motor state space, in which the position of the arm is characterized by the pair of angles (θ, φ). Churchland (1986) argues that the necessary transformation $(\theta, \varphi) = f(\alpha, \beta)$ is complicated, but it can be learned in principle by coupling the sensory and motor maps (Fig. 4.2).

R. Engbert, *Dynamical Models In Neurocognitive Psychology*,
https://doi.org/10.1007/978-3-030-67299-7_4

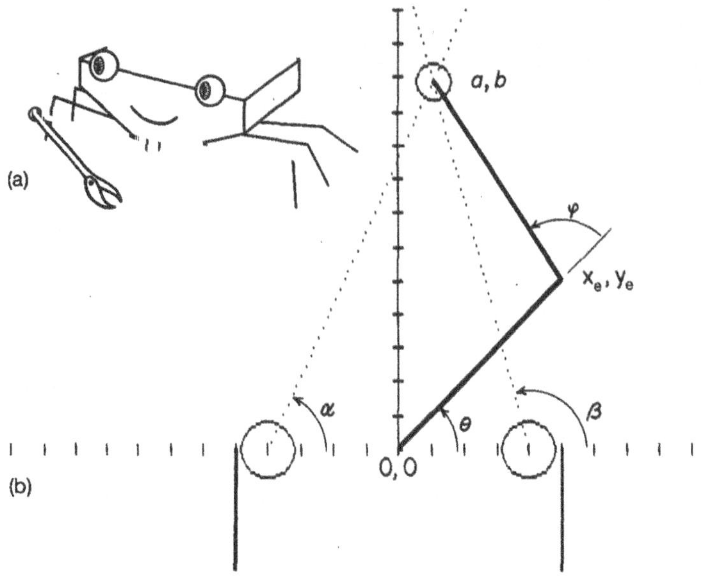

Fig. 4.1 The crab Robin as a model for sensorimotor integration. To reach a target object with the manipulator arm, a relationship between the pairs of angles (α, β) of the eyes and the angles (θ, φ) of the arm must be established (from: [3])

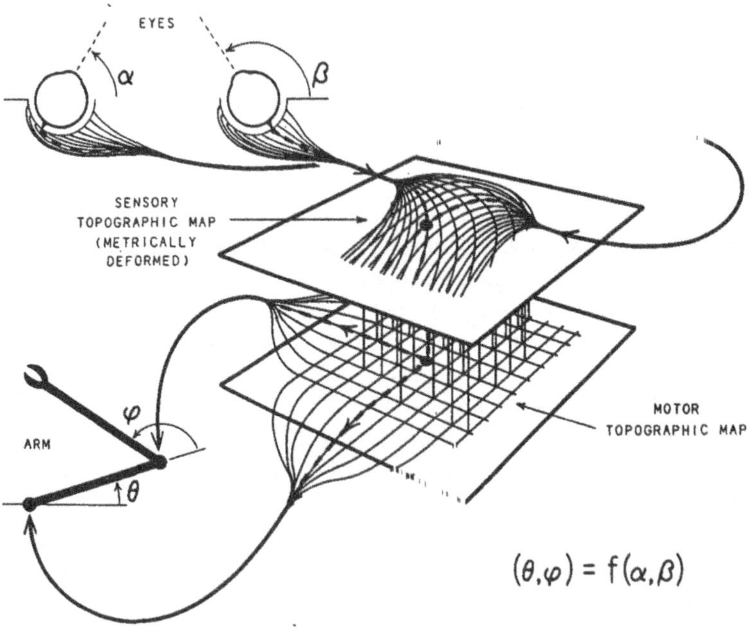

Fig. 4.2 Illustration of the suggested neurophysiological basis of the transformation (θ, φ) = $f(\alpha, \beta)$ from the sensory map of the eyes' angles to the motor map of the manipulator arm (from: [3])

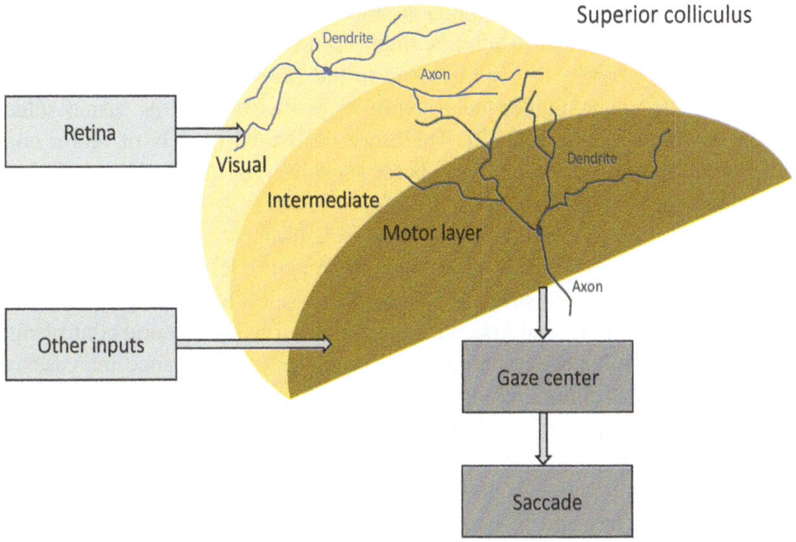

Fig. 4.3 Neurophysiological layers of the superior colliculus (SC). The SC receives signals from the retina, which converge to a sensory layer to define a sensory map. This sensory map is connected to a motor layer that determines the resulting saccade (after: [17])

For the biological implementation of the required transformation from sensory to motor space, the physiological structure of the *superior colliculus* provides a suitable anatomy, since it combines two separated sensory and motor layers (Fig. 4.3). In the sensory layer, visual neurons could be identified that receive input from retinal ganglion cells. Axons of these visual neurons in the sensory layer project into the ventral motor layer, which controls saccadic eye movements as discussed in Sect. 2.1. Thus, given some plasticity of the coupling between sensory and motor layers, the neurophysiology of saccade generation is highly compatible with Churchland's concept of a (learned) transformation between sensory and motor maps (Fig. 4.2). For the eyes, such a transformation is possible because the eyes do not carry any load, while other motor systems (e.g., the hands) are typically used for object manipulation. Thus, stereotyped movements resulting from Churchland's transformation principles do not necessarily generalize to other motor systems, in particular, effectors for object manipulation.

During the last decade, the analysis of statistical fluctuations in sensorimotor processes has become more influential. Both sensory and motor processes are subject to noise. In the case of sensory systems, we have to learn statistical properties of our natural environment. Natural scenes are complex, but they share inherent statistical regularities. For example, Mandelbrot (1975, 1983) published a series of important papers on the analysis of *statistical self-similarity* [14, 15]. As an example, the power spectral density of the spatial frequency content in natural scenes can be described by a power-law $1/f^{\beta}$, where the exponent β is approximately $\beta \approx 2$ [6, 19, 20].

Because of stochastic fluctuations in the neural coding of stimuli, we also have to learn the statistical properties of the sensory system itself. For example, vision is characterized by ambiguities due to the 2D retinal representations of 3D stimuli [21]. Moreover, higher-level cognitive processes such as visual selective attention modulate the percept [2]. Therefore, statistical models of motor control may be looked upon as appropriate tools for sensorimotor integration under various uncertainties.

In addition to the sensory systems, motor execution processes represent another source of noise. As a general principle, the magnitude of the noise in motor systems is correlated to the strength of the signal—an effect that is termed *signal-dependent noise* [7]. In this approach it has been observed that neural control signals are typically contaminated by neural noise, such that the variance increases with the amplitude of the command signal. In this situation, it can be shown [7] that the properties of a trajectory are determined by minimization of the variance of the resulting movement.

Ironically, a robotic effector with the magnitude of noise found in human motor systems could not be used in current robotics due to the lack of robust control principles. From this perspective, the shift of the research strategy in motor control to probabilistic models seems to be important for future success in the control of artificial devices for movements and object manipulation. Bayesian models discussed in this chapter represent a very important class of these probabilistic models.

4.2 Bayesian Motor Planning

Our sensory systems do not have direct access to the parameters specifying the physical properties of the surrounding environment. Using cognitive processing of sensory signals, our nervous system has to rely on statistical inference about parameters θ based on neural data s. This process can be formulated as *maximum likelihood estimation* (MLE) that determines the parameter values θ that maximize the observation of given neural data s. Formally, we can define the *likelihood function* $Q(s|\theta)$, which quantifies the probability of the neural data s given the true parameter values θ.

If one asks for the specific motor action that is optimally adapted to a given stimulus, then our primary interest is focused on the parameters θ. Which values of the set of parameters θ are (most probably) responsible for the generation of the observed neural data s? Using the *Bayes' Theorem* (see Sect. A.1.8) we can write down the probability of the parameters θ after observing the neural data s as

$$\underbrace{P(\theta|s)}_{\text{Posterior}} = \frac{\overbrace{Q(s|\theta)}^{\text{Likelihood}} \times \overbrace{P(\theta)}^{\text{Prior}}}{R(s)}, \tag{4.1}$$

where the *prior probability* $P(\theta)$ describes our knowledge on the probability density of the parameters before the observation of the neural data s and $R(s)$ is the probability density of all possible observational data. It is important to note that knowledge of $R(s)$ is not required, since we are interested in the functional dependence of the posterior from θ. The factor $R(s)$, however, does not change this functional form, thus it can be interpreted as a normalization constant of the posterior. The maximum of the *posterior probability* $P(\theta|s)$ after the observation of the data s represents the optimal estimate of the parameters θ.

The theorem by Thomas Bayes (1701–1761), Eq. (4.1), was published posthumously in 1763 [1]. It represents a consistency relation for conditional probabilities, which directly follows from the definition of conditional probabilities, i.e.,

$$P(\theta, s) = P(\theta|s)R(s) = Q(s|\theta)P(\theta) . \tag{4.2}$$

This equation can be used to calculate the (un-normalized) posterior probability as

$$P(\theta|s) \propto P(\theta)Q(s|\theta) \tag{4.3}$$

with the constant of proportionality (or normalization constant) given by

$$R(s) = \int_\theta Q(s|\theta)P(\theta)\mathrm{d}\theta . \tag{4.4}$$

Since probability densities are normalized by definition, it is sufficient to determine the form of the probability. As a consequence, the normalization constant does not need to be specified explicitly in most calculations.

The application of Bayes' Theorem to motor planning is illustrated in Fig. 4.4. A participant is asked to estimate the horizontal position of a target object (e.g., a cursor on a display). The prior probability $P(\theta)$ represents the distribution of positions of the cursor that have been experienced during previous trials. It is important to note that this prior knowledge enables an observer to generate a probabilistic prediction of future sensations, which is independent of the sensation of the upcoming event. If the cursor is presented at the horizontal position θ =+1 cm in the next trial, then neural data s are generated in the visual system of the observer with likelihood $Q(s|\theta)$. Using Bayes' Theorem for the given problem, Eq. (4.2), we can calculate the posterior probability $P(\theta|s)$ as a product of the form given in Eq. (4.3). The result is illustrated in Fig. 4.4. In the case of unimodal distributions, the maximum of the posterior probability is located between the maximum of the likelihood and the maximum of the prior probability.

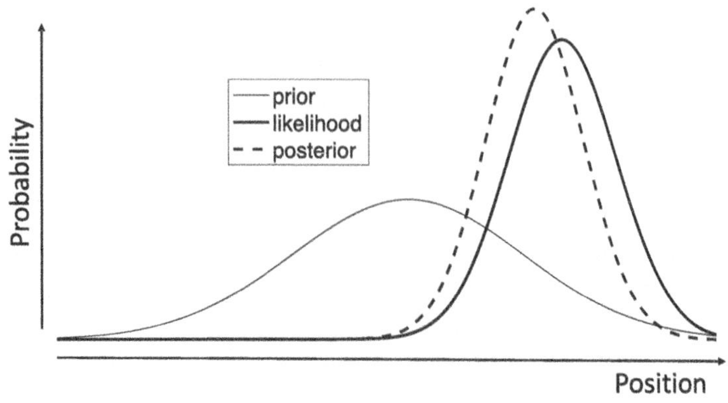

Fig. 4.4 Bayesian motor planning. The task-dependent prior probability is multiplied by a sensory likelihood to obtain a posterior probability that is the basis for optimal decisions in movement preparation (after: [22])

The results in Fig. 4.4 illustrate a general question for the psychological interpretation of the Bayesian approach to motor planning: Why should a systematically biased posterior probability be an optimal estimate for motor planning? This answer is related to the mean-square deviation of the movement target. The multiplication of prior probability and likelihood function generates a posterior probability that is characterized by a reduction in variance. As a result, the Bayesian estimate of the target position is automatically more precise than the sensation (likelihood estimate), however, at the cost of generating a small bias.

In the next section, we will use the Bayesian approach discussed so far for an application to eye guidance in reading, where saccadic landing positions within words will be analyzed.

4.3 Application: Saccades During Reading

Closely related to the topic of the next chapter, we now investigate a part of the problem of eye guidance during reading. Assume that a target word for the next saccade has been selected by cognitive processing, how precise will be the upcoming saccade to the selected target word?

Because of the high receptor density in the fovea, centrally fixated words are processed most efficiently. Therefore, a straightforward assumption is that the word center is the functional saccade target. However, within-word landing positions are typically rather broad, which is shown in Fig. 4.5. The experimentally observed standard deviation of the landing position calls for an explanation, since basic oculomotor paradigms show that the mean-square error in saccade targeting can be much smaller [8] than the error that is found in reading experiments.

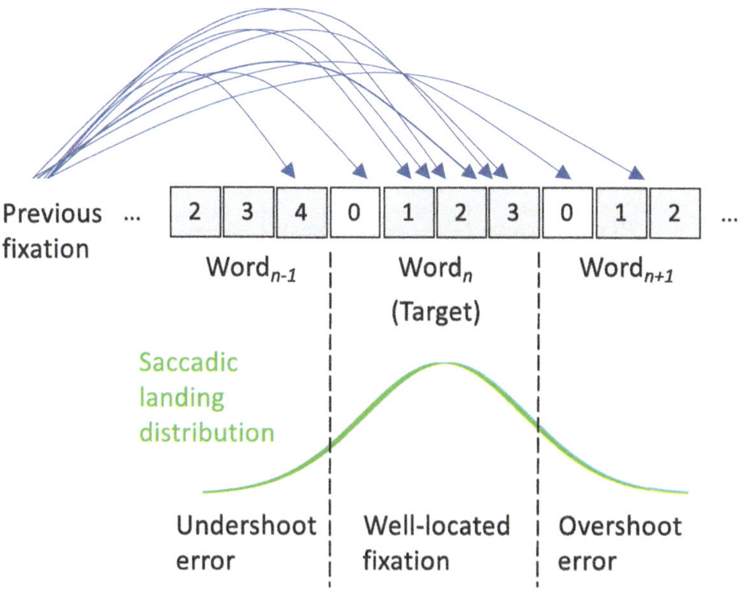

Fig. 4.5 Distribution of saccadic landing positions during reading. From a given fixation position (*launch site*), the saccade is programmed towards the center of the target word. Because of the rather broad distribution of landing positions, the target word is frequently missed due to short (*undershoot*) or long (*overshoot*) saccades (after: [5])

An important consequence of the broad landing-position distribution of saccades in reading is the existence of *mislocated fixations*, where saccadic errors lead to a fixation on a neighboring word of the intended target. Such mislocated fixations have been predicted by McConkie et al. [16]. A first quantitative investigation of mislocated fixation was based on statistical computations of the overlapping landing-position distributions [5], where the computations exploited the fact that proportions of mislocated fixations need to be self-consistent with the saccadic errors underlying the landing-position distributions.

The Bayesian principle of motor planning in reading can be calculated analytically, if we describe both prior probability and likelihood function by normal distributions. For the prior probability, we assume a Gaussian distribution with mean value μ_t and variance σ_t^2,

$$P(x) = \mathcal{N}(x; \mu_t, \sigma_t^2) \; . \tag{4.5}$$

The (sensory) likelihood is the conditional probability for a sensory estimation of the target at position x_o, if the true target position (the word center of the selected target word) is located at position x,

$$Q(x_o|x) = \mathcal{N}(x; x_o, \sigma_o^2) \; . \tag{4.6}$$

The mean value of the likelihood is given by x_o and the variance is denoted by σ_o^2. Thus, we assume that sensory perception is unbiased. With the assumptions on prior probability and likelihood, Eqs. (4.5) and (4.6), the Bayesian model of saccade planning is complete. Multiplication of both probabilities, Eq. (4.3), yields the posterior probability, i.e.,

$$P(x|x_o) \propto \exp\left\{-\frac{(x-\mu_t)^2}{2\sigma_t^2}\right\} \exp\left\{-\frac{(x-x_o)^2}{2\sigma_o^2}\right\} \tag{4.7}$$

$$\propto \exp\left\{-\frac{(x-\mu_p)^2}{2\sigma_p^2}\right\} . \tag{4.8}$$

The resulting probability density can be rearranged in the form of a normal distribution, $P(x|x_o) = \mathcal{N}(x; x_p, \sigma_p^2)$, with parameters

$$\mu_p = \frac{\sigma_o^2 \mu_t + \sigma_t^2 x_o}{\sigma_o^2 + \sigma_t^2} \quad \text{and} \tag{4.9}$$

$$\sigma_p^2 = \frac{\sigma_o^2 \sigma_t^2}{\sigma_o^2 + \sigma_t^2} . \tag{4.10}$$

For the mean value, we can distinguish two extreme cases, which represent, on the one hand, the case of maximum sensory uncertainty ($\sigma_o^2 \rightarrow \infty$) and, on the other hand, minimum sensory uncertainty ($\sigma_o^2 \rightarrow 0$), which give the following predictions:

$$\mu_p \rightarrow \begin{cases} \mu_t \ , \ \text{if } \sigma_o^2 \rightarrow \infty \text{ (max. sensory uncertainty)} \\ x_o \ , \ \text{if } \sigma_o^2 \rightarrow 0 \ \ \text{(vanishing sensory error)} \end{cases} . \tag{4.11}$$

Both limiting cases can be interpreted in a psychologically meaningful way: Without any sensation (or, equivalently, with infinite uncertainty of the sensory percept), a reader might fully rely on prior knowledge of the expected target position from earlier experience and generate a random saccade from the prior probability. In contrast, in a situation of high-accuracy sensory information, the reader would probably use sensory information for the computation of the target position while neglecting prior knowledge.

Experimental data on within-word landing positions indicate a systematic shift of the mean landing position in the direction of the launch-site position (Fig. 4.6). In the Bayesian model, this effect is reproduced by a shift of the position of the maximum of the posterior probability (Fig. 4.7). For experimental studies, the prediction is that saccades will be too short in the case of large launch-site distances, which produce, on average, an *undershoot* of the target position (Fig. 4.7a). In the case of a short launch-site distance, a saccade will be long and generate an *overshoot* error of the target word's center (Fig. 4.7b), on average.

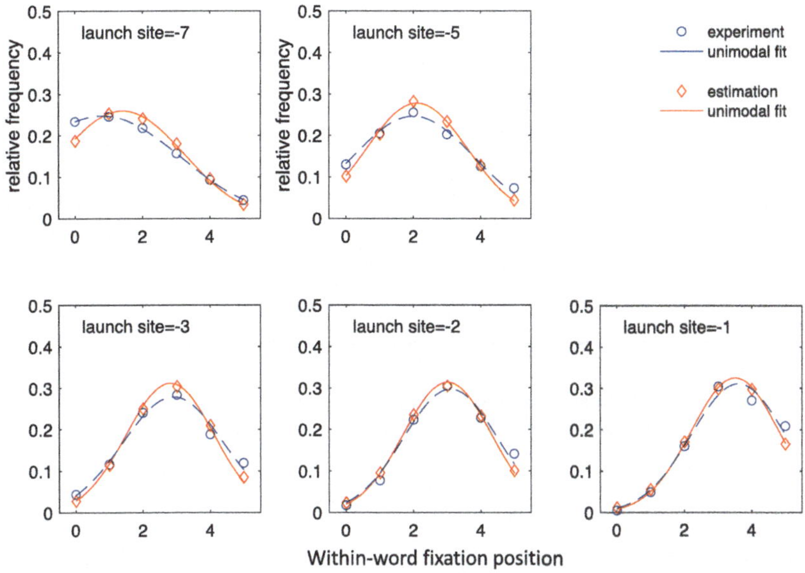

Fig. 4.6 Experimentally observed landing-position distributions. The data show that mean landing position is systematically shifted to the left with increasing distance of the launch site [4]

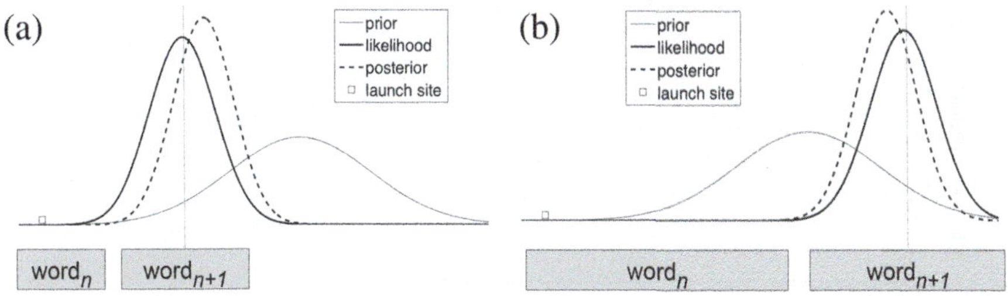

Fig. 4.7 Systematic shift of the mean landing position in the Bayesian model. (**a**) At a large *launch-site distance* a mean *undershoot* error is observed. (**b**) For shorter distance from the launch site to the target word, an *overshoot* is observed on average (after: [4])

The *launch-site effect* on the landing position was discovered by McConkie and colleagues in 1988 [16]. To visualize the main effect, the centered mean landing-position was plotted as a function of the center-based launch-site position (Fig. 4.8a). As a result, a strong linear relation between both variables is obtained. This effect is more or less independent of word length.

In the Bayesian model, the launch-site effect can be analyzed quantitatively. The mean saccadic error, i.e., the mean deviation of the saccadic landing position from the word center, is given as

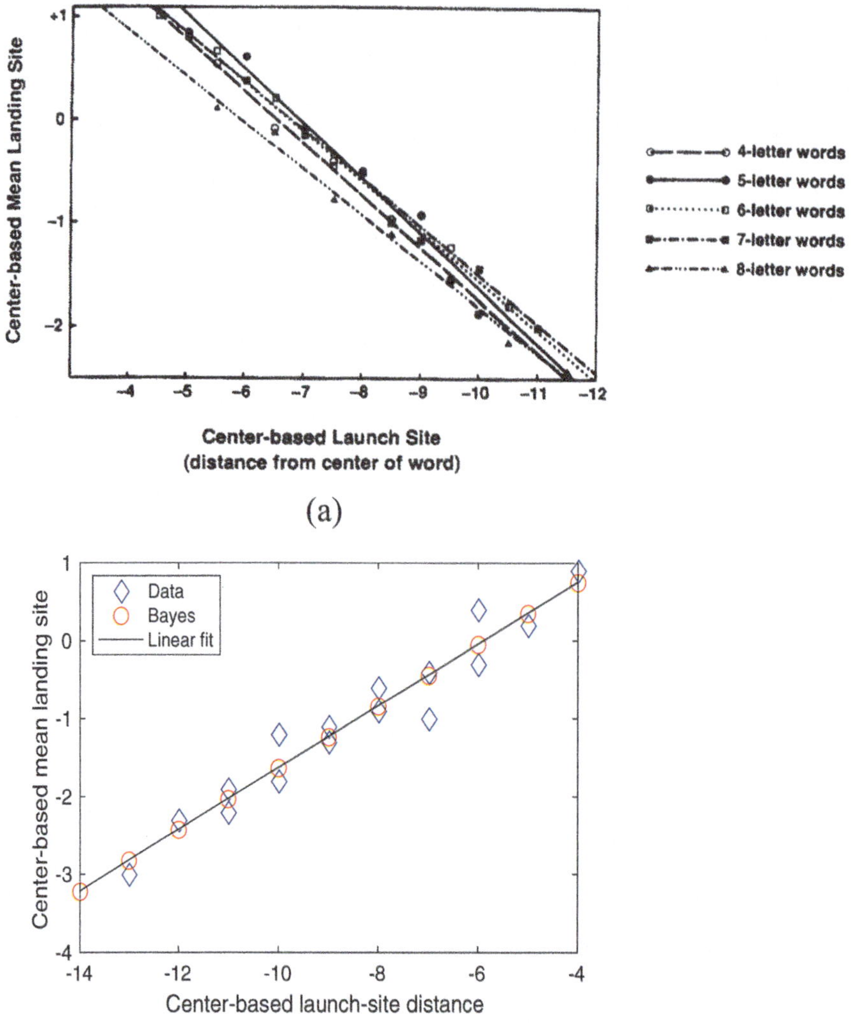

Fig. 4.8 The launch-site effect of saccadic landing positions. (**a**) Original plot the deviation of the mean landing position from the word center as a function of the launch-site distance to the word center [16]. Note the inverted horizontal axis. (**b**) Plot for experimental data obtained from the Potsdam Sentence Corpus [9]. The Bayesian model provides a good fit of the data

$$\Delta_{\text{sacc}} = \mu_P - x_o = \frac{\sigma_o^2}{\sigma_o^2 + \sigma_t^2}(\mu_t - x_o) , \tag{4.12}$$

which can reproduce the experimentally observed relationship [16] very precisely. Using the numerical value of the slope of the linear regression of the data,

$$\frac{\sigma_o^2}{\sigma_o^2 + \sigma_t^2} \approx 0.5 , \tag{4.13}$$

we obtain the prediction that the variance of the sensory uncertainty is approximately the same as the variance of the prior probability, i.e., $\sigma_o^2 \approx \sigma_t^2$.

With the help of numerical simulations (after parameter identification), the Bayesian model was compared to experimental data [4]. Such simulations indicated that the Bayesian model is in good agreement with experimental data of the launch-site effect in reading. It could also be shown that the Bayesian model performs significantly better than a model with random saccade length, i.e., random sampling from the distribution of the experimentally observed saccade lengths. Moreover, another alternative model postulating that saccades target the beginnings of words in case of long launch-site distances and word endings in case of short launch-site distances was investigated. This model performed only slightly better than the random saccade model.

The Bayesian model of with-word target selection is based on the assumption that a cognitive mechanism exists for the selection of target words. Given a selected target word, the processes of sensorimotor integration discussed in this chapter will compute the saccade parameters to generate the required gaze shift for optimal processing of the target word. In the next chapter, we will present the two most influential cognitive models of eye-movement control in reading. Both models are compatible with the Bayesian principles of sensorimotor integration.

4.4 Exercises

Exercise 4.1 (Sensorimotor Transformation in the Crab Robin) Find the functional relation $(a, b) = f(\alpha, \beta)$ for the transformation of the eye's rotation angles α and β to the object's position (a, b) (Fig. 4.2). Assume that the distance of the eyes is given as $2l$.

Exercise 4.2 (Bayesian Analysis of Word Targeting in Reading) Show via analytical calculations that the mean, Eq. (4.9), and standard deviation, Eq. (4.10), of the posterior probability can be obtained via rearranging Eq. (4.7) by completing the square.

Exercise 4.3 (Analysis of the Launch-Site Effect in Reading) Investigate the data on within-word landing positions. (a) Use Gaussian fits of the landing-position distributions to estimate mean value and standard deviation for each combination of launch-site distance and word length. (b) Reproduce the plot of the launch-site effect (Fig. 4.8b).

4.5 Further Reading

The first experimental evidence for Bayesian motor planning was published in 2004 by Körding and Wolpert [10]. For the selection of optimal decisions in the context of motor control, the authors extended their model to Bayesian decision theory in motor control [11]. In the case of saccades during reading, Krügel and Engbert (2014) revised their original Bayesian model [4] to include explicit computations of word centers from word boundaries and to model the case of skipping saccades [13]. An important effect relevant to the analysis of within-word fixation positions is due to mislocated fixations, introduced by [16] and quantitatively studied by [5] and [12].

References

1. Bayes, T. (1763). An essay towards solving a problem in the doctrine of chances. *Philosophical Transactions (1683–1775)*, 370–418. http://www.jstor.org/stable/105741
2. Carrasco, M., Ling, S., & Read, S. (2004). Attention alters appearance. *Nature Neuroscience, 7*(3), 308–313. https://doi.org/10.1038/nn1194
3. Churchland, P. M. (1989). Some reductive strategies in cognitive neurobiology. In S. Silvers (Ed.), *Rerepresentation: Readings in the philosophy of mental representation* (pp. 223–253). Dordrecht, Netherlands: Springer. https://doi.org/10.1007/978-94-009-2649-3_12.
4. Engbert, R., & Krügel, A. (2010). Readers use Bayesian estimation for eye movement control. *Psychological Science, 21*(3), 366–371. https://doi.org/10.1177/0956797610362060
5. Engbert, R., & Nuthmann, A. (2008). Self-consistent estimation of mislocated fixations during reading. *PLoS One, 3*(2), e1534, 1–6. https://doi.org/10.1371/journal.pone.0001534
6. Field, D. J. (1987). Relations between the statistics of natural images and the response properties of cortical cells. *Journal of the Optical Society of America A, 4*(12), 2379–2394. https://doi.org/10.1364/JOSAA.4.002379
7. Harris, C. M., & Wolpert, D. M. (1998). Signal-dependent noise determines motor planning. *Nature, 394*(6695), 780–784. https://doi.org/10.1038/29528
8. Kapoula, Z. (1985). Evidence for a range effect in the saccadic system. *Vision Research, 25*(8), 1155–1157. https://doi.org/10.1016/0042-6989(85)90105-1
9. Kliegl, R., Nuthmann, A., & Engbert, R. (2006). Tracking the mind during reading: The influence of past, present, and future words on fixation durations. *Journal of Experimental Psychology: General, 135*(1), 12–35. https://doi.org/10.1037/0096-3445.135.1.12
10. Körding, K. P., & Wolpert, D. M. (2004). Bayesian integration in sensorimotor learning. *Nature, 427*(6971), 244–247. https://doi.org/10.1038/nature02169
11. Körding, K. P., & Wolpert, D. M. (2006). Bayesian decision theory in sensorimotor control. *Trends in Cognitive Sciences, 10*(7), 319–326. https://doi.org/10.1016/j.tics.2006.05.003
12. Krügel, A., & Engbert, R. (2010). On the launch-site effect for skipped words during reading. *Vision Research, 50*(16), 1532–1539. https://doi.org/10.1016/j.visres.2010.05.009

13. Krügel, A., & Engbert, R. (2014). A model of saccadic landing positions in reading under the influence of sensory noise. *Visual Cognition, 22*(3–4), 334–353. https://doi.org/10.1080/13506285.2014.894166
14. Mandelbrot, B. (1983). *The fractal geometry of nature* (495 p.). New York: WH Freeman and Co. https://us.macmillan.com/thefractalgeometryofnature/benoitbmandelbrot/9780716711865
15. Mandelbrot, B. B. (1975). Stochastic models for the earth's relief, the shape and the fractal dimension of the coastlines, and the number-area rule for islands. *Proceedings of the National Academy of Sciences of the United States of America, 72*(10), 3825–3828. http://www.pnas.org/content/72/10/3825
16. McConkie, G. W., Kerr, P. W., Reddix, M. D., & Zola, D. (1988). Eye movement control during reading: I. The location of initial eye fixations on words. *Vision Research, 28*(10), 1107–1118. https://doi.org/10.1016/0042-6989(88)90137-X
17. Purves, D., Augustine, G. J., Fitzpatrick, D., Hall, W. C., LaMantia, A.-S., & White, L. E. (2012). *Neuroscience* (5th ed.). Sunderland/MA: Sinauer Associates Inc. https://www.sinauer.com/neuroscience-770.html
18. Rosenbaum, D. A. (2009). *Human motor control*. Cambridge: Academic Press. http://psycnet.apa.org/psycinfo/2010-06119-000
19. Simoncelli, E. P., & Olshausen, B. A. (2001). Natural image statistics and neural representation. *Annual Review of Neuroscience, 24*(1), 1193–1216. https://doi.org/10.1146/annurev.neuro.24.1.1193
20. van der Schaaf, V. A., & van Hateren, J. H. V. (1996). Modelling the power spectra of natural images: Statistics and information. *Vision Research, 36*(17), 2759–2770. https://doi.org/10.1016/0042-6989(96)00002-8
21. Wolfe, J. M., Kluender, K. R., Levi, D. M., Bartoshuk, L. M., & Herz, R. S. (2015). *Sensation and perception* (4th ed.). Sunderland/MA: Sinauer Associates Inc. https://www.sinauer.com/sensation-perception-784.html
22. Wolpert, D. M. (2007). Probabilistic models in human sensorimotor control. *Human Movement Science, 26*(4), 511–524. https://doi.org/10.1016/j.humov.2007.05.005

Chapter 5
Eye-Movement Control During Reading

The general function of eye movements is to shift gaze position from one patch (or object) within a given visual scene to another. Thus, eye movements represent the active part of visual information processing, often termed *active vision* [12]. Because of the complexity of natural scenes, the control of gaze position is influenced by a large number of variables and processes. In reading, eye guidance can be studied within a well-structured, effectively one-dimensional spatial environment (i.e., the line of text). Since reading is based on the coordination of several key human cognitive subsystems (e.g., vision, attention, memory, eye movements), computational models are needed to generate predictions from theories integrating these subsystems in a dynamical framework. Consequently, mathematical models of active vision play a crucial role in driving the research questions in the research field of eye movements during reading (see [27] for a recent discussion).

In the previous chapter, we investigated the problem of saccade planning during reading and developed a mathematical model of oculomotor control given that target words were selected by visual-cognitive processes that were not further specified. Assuming a selected target word, we proposed a Bayesian model for sensorimotor integration, which reproduced the launch-site effect of within-word fixation position. Here we discuss cognitive models that were developed for the selection of target words during reading.

Traditional theories in cognitive psychology are based on the assumption of a sequential-modular organization of cognition and behavior. Specifically, cyclic sequences of activity in the form of perception $\mapsto$ thinking/decision $\mapsto$ action $\mapsto$ perception $\mapsto$ ... are generated during ongoing behavior. However, as we have seen in the example of fixational eye movements and microsaccades, a unidirectional coupling of perception and action is implausible, since visual perception is both the purpose of and the basis for eye movements. More recent theories question the classical sequential organization of perception, decision, and action. In the alternative framework, the *dynamical hypothesis* of cognition [34], different cognitive and physiological processes can interact at any time based on a set of time-dependent

R. Engbert, *Dynamical Models In Neurocognitive Psychology*,
https://doi.org/10.1007/978-3-030-67299-7_5

rules that are very similar to equations of motions [4] in the physical sciences (e.g., Newton's second law, Navier-Stokes equations).

5.1 Eye Movements in Reading

Reading is a fundamental cognitive skill that demonstrates how efficiently different cognitive processes can be coordinated to generate adaptive behavior. The key cognitive subsystems involved in eye-movement control during reading are visual perception and information processing, word recognition, memory retrieval, selective attention, and motor planning. An important research goal is to derive the critical set of rules for interaction of these subsystems to build a process-oriented model of reading.

Even during reading of simple texts, eye movements produce a complicated sequence of fixation (a *scanpath*), since fixations of words are typically not in perfect serial order (Fig. 5.1). Only about 50% of the saccades move the gaze from word_n to word_{n+1}. *Refixations* represent a fraction of roughly 20% of the saccades, which relocate the fixation position within the same word. These refixations occur most often when reading long or low-frequency words. A saccade from word_n to word_{n+2} is denoted as *word skipping*. Word skipping saccades are also frequent with a proportion of 20%, where the numerical value depends on text difficulty, reading skill, and intention of the reader, of course. For the analysis of eye movements in reading, *regressions* are particularly challenging, since regressions move the gaze against reading direction to parts of text that were previously visited. Regressions represent about 10% of all saccades (Fig. 5.2).

A central question for models of eye-movement control in reading is whether the non-serial sequence of fixations typically observed in reading experiments is generated because of underlying non-serial word processing. On the level of theories, the assumptions put forward by different models [30] include both extreme cases of strictly serial word processing and of parallel lexical processing of words. In the *E-Z Reader* model [29], the authors assume that word order is strictly preserved during processing. Due to partial uncoupling between fixation position and attention allocation, however, the E-Z Reader can explain the four different saccade types

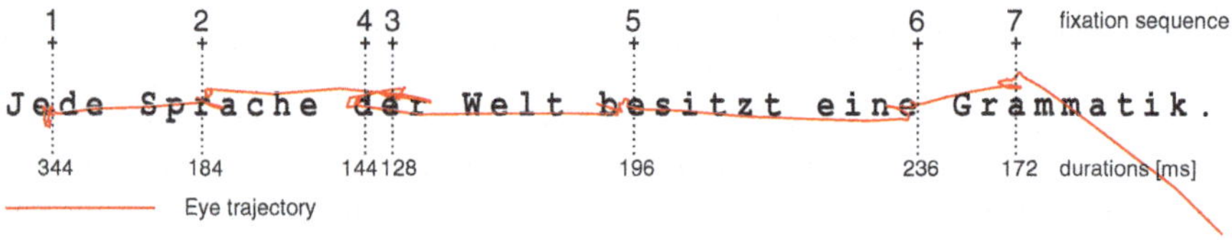

Fig. 5.1 Example of the eye's scanpath during reading. Horizontal fixation positions are indicated by the dashed vertical lines, numbered by the upper labels. Fixation durations in milliseconds are given by the lower labels

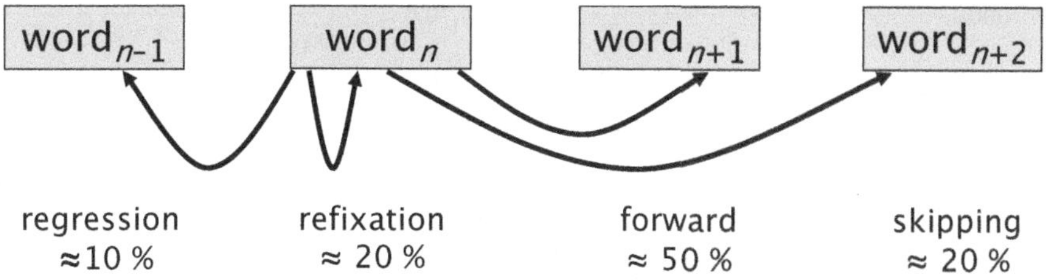

Fig. 5.2 Saccade types and relative frequencies during reading. About half of all saccades move the foveal part of the visual field from word$_n$ to the next word$_{n+1}$. Other saccade types generate *refixations*, *word skipping*, and *regressions*

(Fig. 5.2) in principle. Moreover, *mislocated fixations* have been suggested as an explanation of seemingly non-serial processing [5], however, such an explanation has not been investigated quantitatively (and statistical modeling of mislocated fixations [9] has been neglected so far).

In the *SWIFT* model [8, 10], it is assumed that several words can be processed in parallel, i.e., processing for neighboring words is partially overlapping. As a result, word order is not necessarily preserved during cognitive processing. The four saccade types are generated from a dynamical activation field that determines the probabilities for saccadic target selection.

Besides different saccade types, reading experiments generate rich data on fixation durations and fixation probabilities. Therefore, a range of quantitative measures has been established to characterize reading behavior. Because of the complicated scanpath or fixation sequence, these measures are typically conditional. For example, the *single-fixation duration* gives the duration of the fixation of a particular word if it is fixated only once. Additionally, the *first-fixation duration* is the duration of the first fixation if the word is fixated two or more times in a direct sequence. All fixation that occur before the first regression are denoted as *first-pass*. The *gaze duration* is the sum of the durations of all fixations (Fig. 5.3a, c) in first-pass, while *total viewing time* measures the sum of all fixation durations independently of the realized sequence of fixations in a given trial. At first sight, these measures appear to be arbitrarily defined. However, experimental research over the last 50 years demonstrated that the above range of fixation-duration measure summarizes reading behavior in a meaningful way.

For fixation probabilities, conditional measures include the *skipping probability*, which is the probability of skipping a particular word during first-pass reading. Other measures are the probabilities for two fixations or three and more fixations of a word in first-pass. Since regressions play an important role in theoretical models, the probabilities that a certain word is the target of a regression, i.e., the *regression probability*, is an additional measure that reflects the complexity of the fixation sequence (Fig. 5.3b, d).

Both fixation durations and fixation probabilities show highly informative dependencies on statistical predictors like printed word frequency (Fig. 5.3a, b) or word

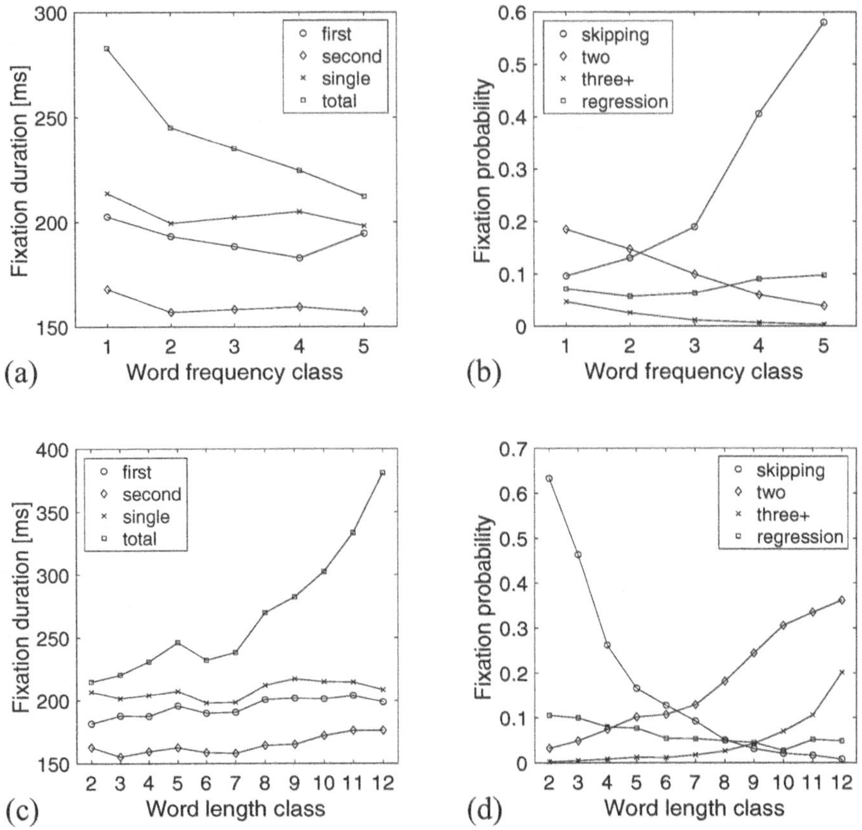

Fig. 5.3 Summary statistics of fixation durations and fixation probabilities during reading [10]. (**a**) Fixation duration as a function of word frequency. (**b**) Fixation probability as a function of word frequency. (**c**) Fixation duration as a function of word length. (**d**) Fixation probability as a function of word length

length (Fig. 5.3c, d). Word frequency is often looked upon as the most important predictor that characterizes cognitive processing difficulty. It is estimated from printed word frequency by counting the relative numbers of occurrences of a given word in large text corpora [17].

For model building, the dependence of quantitative measures of eye-movement behavior on statistical predictor variables and their interactions provide important boundary conditions [18]. Additional effects are even more specific. For example, mean fixation duration is systematically shorter before skipped words compared to fixation duration before fixated words [16]. Another example is that mean fixation durations are shorter if the fixation position is close to the word's edges compared to the word's center [24]. In the next sections, we will discuss different model assumptions that aim at a theoretical understanding of eye-movement control during reading. Each of the models makes highly specific predictions, which can, however, only be obtained from numerical simulations.

5.2 A Serial Two-State Model

We start with a very simple model of eye-movement control during reading by assuming that two sub-processes, (1) lexical processing of words (word recognition) and (2) programming of saccades, are operating sequentially. If we denote lexical processing of word$_n$ by l_n and programming of a saccade to word$_{n+1}$ by s_{n+1}, then we can represent the functioning of such a serial model by the sequence

$$l_1 \rightarrow s_2 \rightarrow l_2 \rightarrow s_3 \rightarrow l_3 \rightarrow s_4 \rightarrow l_4 \rightarrow s_5 \quad \ldots, \tag{5.1}$$

which illustrates that the model switches between the two *internal states*, (1)="l" and (2)="s". The interaction of these two sub-processes can be illustrated graphically by the internal states and its transitions (Fig. 5.4).

Beyond its two-state architecture, the serial model needs to be completed by specifications of the two sub-processes, lexical access, l_n, and saccade programming, s_n. First, saccades are very fast (ballistic) eye movements that cannot be adjusted during movement executions. As a consequence, all details of the movements must be pre-programmed in advance, a computationally costly process. The resulting mean saccade programming time has been estimated experimentally in oculomotor research [2]. For the simulation of the simple two-state model, a mean duration of 100 ms to 150 ms can be used as a plausible numerical value of the saccade programming time.

Second, lexical processing time of a word is mainly determined by its frequency. As a first approximation, we can assume that mean lexical access time depends on word frequency via a logarithmic relation, i.e.,

$$\langle l_n \rangle = \alpha - \beta \log(F_n), \tag{5.2}$$

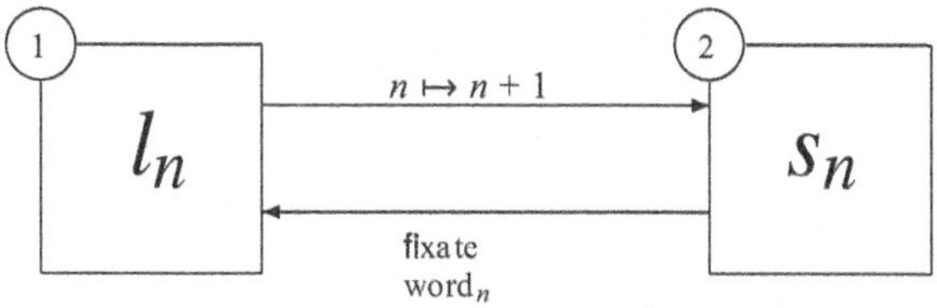

Fig. 5.4 The serial two-state model for the control of eye movements during reading. In state 1, word$_n$ is fixated and lexically processed (denoted by l_n). In state 2, a saccade to word$_n$ is programmed, labeled by s_n Note that the counter variable n is incremented by one ($n \mapsto n+1$) during the transition from state 1 to 2

where F_n is the frequency of word$_n$ and α, β are constant parameters for a given text corpus.

The most important problem of the two-state model is its inefficiency, since word processing is paused during the preparation of eye movements. As a consequence, the rate of word processing of the two-state model is very low. As a solution to this problem, our visual systems exploit information outside the center of the visual field for more efficient reading.

Information processing is most efficient in the central 2° of the visual field, the *foveal* region. Acuity decreases in the *parafoveal* region, which extends out to 5°, and is even poorer in the peripheral region beyond the parafovea. Since saccades are costly, it will turn out to be a considerable advantage, if parafoveal information is used during reading. For processing the word to the right of a currently fixated word, attention has to shift to the parafovea—away from the currently fixated word and to the next word. In this case, programming of a saccade to the next word and lexical access of the currently fixated word are active simultaneously. Therefore, models with attention shifts are characterized by parallel processing of lexical access and programming of saccades, a concept first introduced by Morrison [23]. Regardless of the number of internal states, the basic mechanisms that are performed by these models are shifts of attention and eye movements. The corresponding class of model is often called *sequential attention shifts* (SAS) models. The most important example of this class of models is the E-Z Reader model discussed in the next section.

5.3 The E-Z Reader Model

The *E-Z Reader* model [29] is the most advanced model that operates via sequential attention shifts. From a mathematical point of view, the E-Z Reader model represents a refined version of the basic SAS architecture, with an increased number of internal states and more detailed modeling of sub-processes in lexical access, attention shifts, and saccade programming. In this section, we discuss the simplest version of the model, E-Z Reader 1, to illustrate the modeling strategy in this framework.

The basic assumption of the E-Z Reader model is that there are five sub-processes controlling eye movements during reading [29]: (1) familiarity check of a word, f, (2) completion of lexical access of the word, lc, (3) a labile stage of the saccade program, m, (4) a subsequent non-labile stage of the saccade program, M, and (5) the execution of the saccadic eye movement.

The familiarity check is a sub-process in word recognition, which signals, after termination, that word recognition is imminent. Thus, after a successful familiarity check, a saccade program is started to prepare a saccade to the next word. For a more detailed discussion of the concept of familiarity see Reichle et al. [29]. While the end of the familiarity check triggers a saccade program to the next word, lexical processing of the currently fixated word enters the final stage called lexical

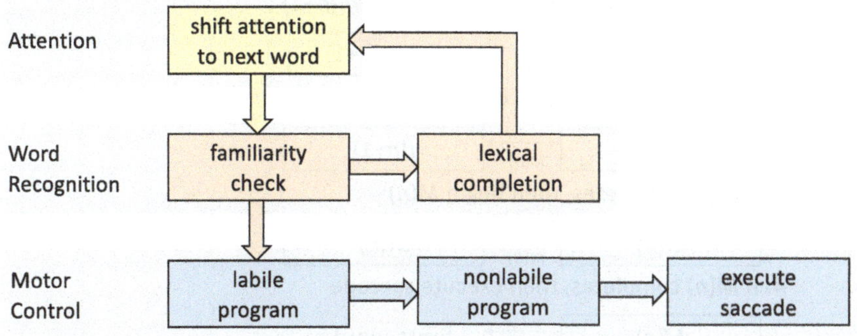

Fig. 5.5 Basic architecture of the E-Z Reader model. Three different modules (attention, word recognition, motor control) are coupled by the familiarity check. After termination of the familiarity check, a saccade program to the next word is initiated and the final stage of lexical processing (lexical completion) starts

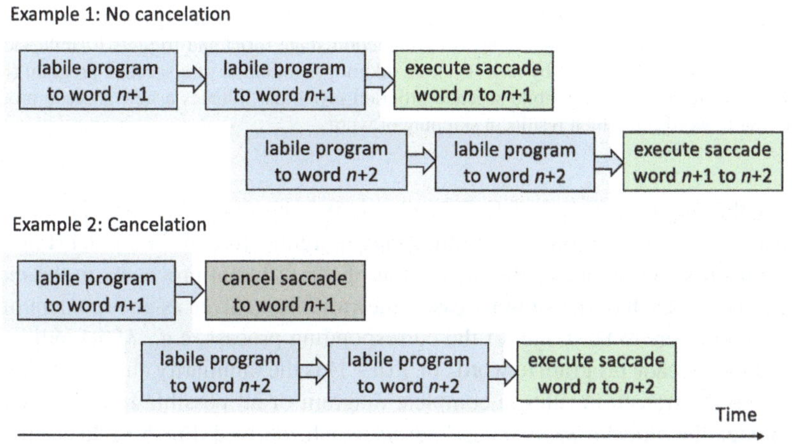

Fig. 5.6 Labile and non-labile stages of the saccade program. Example 1: A second saccade program can be initiated during the non-labile stage of the first program. Example 2: If a second command to start a saccade program occurs during the labile stage of saccade program, the first saccade program is canceled

completion (Fig. 5.5). After lexical completion, attention shifts from the currently fixated word to the next word.

The subdivision of the saccade program into labile and non-labile stages is consistent with experimental results on cancelation and modification of saccade targets [2]. During the labile stage, saccades can still be canceled if a secondary command for saccade programming occurs (Fig. 5.6). After the labile programming stage, however, the saccade can no longer be influenced by other processing events during the non-labile stage.

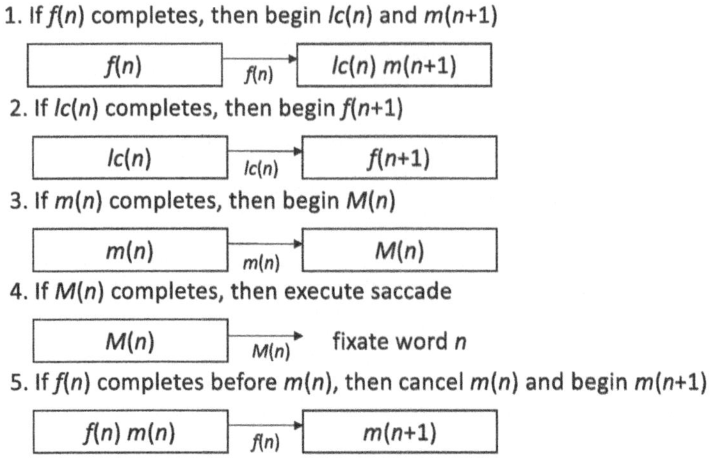

Fig. 5.7 Transition rules of the sub-processes in E-Z Reader 1. The termination of a particular process is indicated by the label of the arrow, e.g., the end of the familiarity check $f(n)$ of word$_n$ (in 1.) leads to the initiation of the lexical completion stage $lc(n)$ and triggers a labile saccade program to word$_{n+1}$, denoted by $m(n+1)$; an additional initiation of a saccade program (in 5.) induces a cancelation of the labile saccade program $m(n)$ and initiates a new saccade program $m(n+1)$ to word$_{n+1}$, which results in skipping of word$_n$

For the implementation of E-Z Reader 1, it is necessary to formulate a set of transition rules in accordance with the general architecture of the model (Fig. 5.5). The transition rules specify the interaction of the different processes in the model (Fig. 5.7). For each of the sub-processes, the variable n (given as an argument of the processes) refers to the target of the corresponding process, e.g., $M(n)$ indicates a non-labile saccade program to word$_n$ or $f(n+1)$ is the familiarity check of word$_{n+1}$.

Using the transition rules, a complete diagram of all possible internal states of the model, the *order-of-processing diagram*, can be derived (Fig. 5.8). In each of the eight internal states, one or more sub-processes are active. If there are several sub-processes in an internal state of the model, then the label of the arrow indicates, which process triggers the corresponding transition to another state, e.g., the transition from state 2 to state 3 is performed, when the non-labile saccade program $M(n+1)$ terminates *before* the lexical completion $lc(n)$ of word$_n$. Alternatively, if $lc(n)$ terminates *before* $M(n+1)$, then a transition to state 7 is generated. The dashed arrow from state 2 to 7 indicates that the index variable n is incremented by 1 to $n+1$.

In E-Z Reader 1, the mean time for lexical access of word$_n$ is a linear function of the logarithm of its frequency F_n. This relation holds for both sub-processes of lexical access, the familiarity check and the lexical completion process (see Eq. (5.2), p. 71). The mean duration of the familiarity check $f(n)$ of word$_n$ is given by

$$\langle f(n)\rangle = f_b - f_m \log(F_n), \tag{5.3}$$

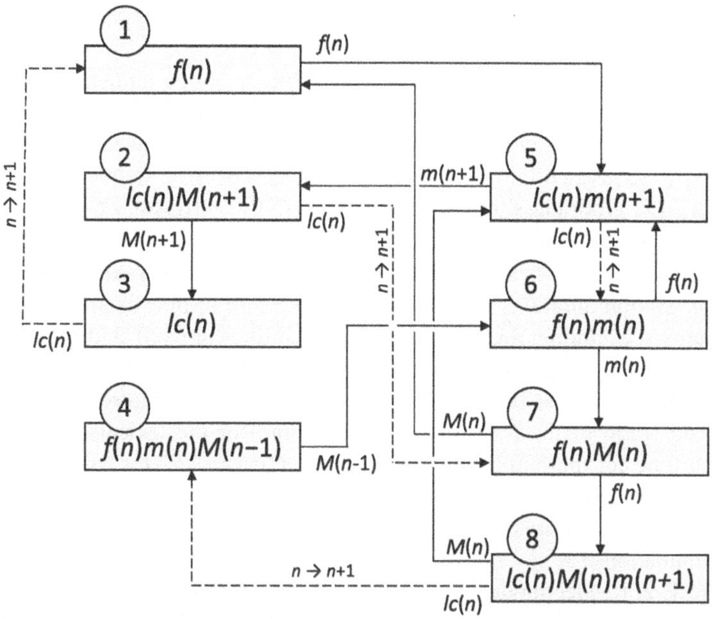

Fig. 5.8 Order-of-processing diagram of the E-Z Reader model. In each of the eight internal states, combinations of different lexical and oculomotor sub-processes can be active

where f_b and f_m are constant parameters and F_n is the word frequency (per million) of word$_n$. Realistic lexical access times fluctuate stochastically around the mean value given by Eq. (5.3). In E-Z Reader, gamma-distributed durations are used for stochastic simulations of the model.

The average duration of lexical completion is assumed to be a fixed proportion Δ of the corresponding duration of the familiarity check [29], i.e.,

$$\langle lc(n) \rangle = \Delta \cdot \langle f(n) \rangle \ . \tag{5.4}$$

How is word skipping generated in E-Z Reader 1? Word skipping occurs, if a labile saccade program $m(.)$ is canceled. As we can see in the OP diagram (Fig. 5.8), such a cancelation occurs during the transition from state 6 to state 5 (see also transition rule 5. in Fig. 5.7). In state 6, the current fixation position is on word$_{n-1}$, while the familiarity check $f(n)$ of word$_n$ is active and a saccade program $m(n)$ to word$_n$ is in preparation. The transition to state 5 is carried out, if familiarity check $f(n)$ is faster than the labile saccade program $m(n)$. Termination of the familiarity check $f(n)$ cancels the saccade program to word$_n$ and continues word processing with lexical completion $lc(n)$. Why is this assumption plausible? Since lexical access of word$_n$ is imminent after termination of the familiarity check, the saccade to word$_n$ is no longer necessary. Thus, the saccade program to word$_n$ is canceled and a new program to word$_{n+1}$ is started. In sum, it is the transition from state 6 to state 5 that determines the skipping of word$_n$.

The E-Z Reader model has been tested quantitatively by numerical simulations of a corpus of 48 sentences [32]. For all 536 words of the corpus, values of word frequency and probability for prediction were available. For the statistical analysis, the words were divided into five logarithmic classes of word frequency (class 1: 1–10; class 2: 11–100; class 3: 101–1000; class 4: 1001–10,000; class 5: >10,000). In the experimental data, all trials including regressions were discarded. For model simulations, 1000 statistical realizations (or model runs) were used to compute mean values of fixation durations (gaze duration = the sum of first-fixation duration and all refixations, excluding regressions) and probabilities of word skipping. Parameter values of the numerical simulations are $f_b = 254$ ms (intercept of the mean completion time of the familiarity check), $f_m = 22$ ms (corresponding slope), $\Delta = 0.65$ (ratio of the length of the mean time for the lexical completion and the mean time for the familiarity check stage), $t(m) = 150$ ms (mean duration of the labile saccade program), and $t(M) = 50$ ms (mean duration of the non-labile saccade program). Results indicate good agreement between experimental data and model simulations [29].

In the E-Z Reader model, waiting times (i.e., stochastic durations of the five sub-processes) were assumed to be gamma-distributed [29]. This assumption has been made to account for the experimentally observed distributions of fixation durations. In the next section, we introduce some general concepts of continuous-time stochastic simulation.

5.4 Numerical Simulation of the E-Z Reader Model

For the simulation of the E-Z Reader model, we can use Gillespie's (1976) *minimal process method* [15, 33]. The master equation is fully determined by the transition probabilities $W_{nn'} \geq 0$. The internal states n are illustrated in the *order-of-processing diagram* (Fig. 5.8).

Starting the simulation in state $n = 1$ at time $t = 0$, i.e., $p_{n'}(t) = 0$ for all $n' \neq n$, we can compute the stochastic waiting time for the first transition to a neighboring state. The total transition probability from state n to all possible adjoined states n' is given by

$$W_n = \sum_{n'} W_{n'n} \, . \tag{5.5}$$

Therefore, we use W_n to compute the exponentially distributed waiting time Δt. In the case of competing transitions to more than one adjoined state, we determine the resulting new state by the relative transition probability.

The simulation algorithm can be formulated as follows:

1. Start in state n at time $t = 0$.
2. Calculate the total transition probability, $W_n(t)$.
3. Determine the waiting time $\Delta t = -\frac{1}{W_n}\ln(1-\xi)$ with the equally distributed random number $0 \leq \xi < 1$.
4. Determine the new state n' according to the relative transition probability, $p_{n'} = w_{n'}/W_n$, using the *linear selection algorithm* (see below).
5. Update state and time in the model, $t \mapsto t + \Delta t$ and $n \mapsto n'$.
6. If $t < T$, continue the simulation with step 1., otherwise terminate the simulation.

An efficient method to choose between N alternatives with different relative probabilities π_j with $(j = 1, 2, \ldots, N)$ is the *linear selection algorithm*:

```
r <- runif(1)
j <- 1
test <- pi[j]
while ( test < r ) {
   j <- j + 1
   test <- test + pi[j]
}
```

The linear selection algorithm guarantees that the stochastic transition j is generated with probability π_j.

5.5 *Analytical Solution of the Master Equation for E-Z Reader

In this paragraph, we compare the numerically simulated stationary probability of the E-Z Reader model with an analytical solution. First, we explain how an analytical solution for the stationary probability distribution can be obtained from the master equation. Second, we compare analytical solution and simulated probability distribution (Fig. 5.9).

The master equation, Eq. (A.29), can be written in the form [33]

$$\frac{\partial}{\partial t} p_n(t) = \sum_{n'} U_{nn'} p_{n'}(t) \tag{5.6}$$

with the definition

$$U_{nn'} := W_{nn'} - \delta_{nn'} \sum_{n''} W_{n''n} \,. \tag{5.7}$$

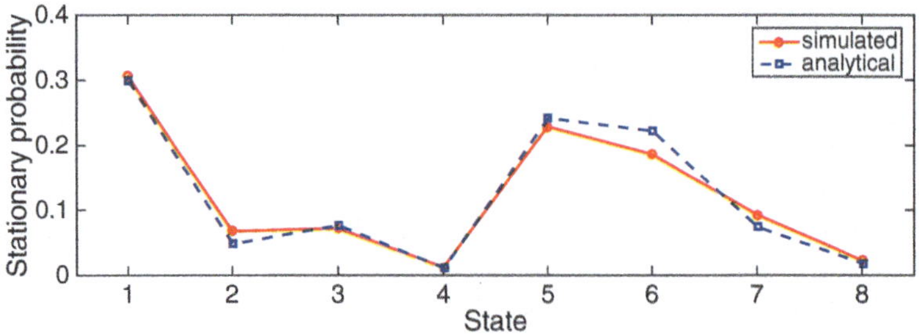

Fig. 5.9 Stationary probability distribution of the E-Z Reader model. Analytical solution, Eq. (5.10), and simulated data are in good agreement

For a stationary solution of a master equation with constant transition probabilities W_{mn} we expect that the system tends to a stationary state for $t \to \infty$, which can be described by the probabilities $\bar{p}_n$ with $\lim_{t\to\infty} p_n(t) = \bar{p}_n$.

For the existence of such stationary solution, the master equation must have at least one non-trivial solution $\bar{p}_n$ with properties

$$\sum_{n'} U_{nn'} \bar{p}_{n'} = 0, \quad \bar{p}_n \geq 0, \quad \sum_n \bar{p}_n = 1\,. \tag{5.8}$$

A more detailed discussion of the solution of the master equation can be found in Schnakenberg [33].

In the E-Z Reader model, the transition probabilities can be written in matrix form as

$$W_{nn'} = \begin{pmatrix} 0 & 0 & \tilde{lc} & 0 & 0 & 0 & \tilde{M} & 0 \\ 0 & 0 & 0 & 0 & \tilde{m} & 0 & 0 & 0 \\ 0 & \tilde{M} & 0 & 0 & 0 & 0 & 0 & 0 \\ 0 & 0 & 0 & 0 & 0 & 0 & 0 & \tilde{lc} \\ \tilde{f} & 0 & 0 & 0 & 0 & \tilde{f} & 0 & \tilde{M} \\ 0 & 0 & 0 & \tilde{M} & \tilde{lc} & 0 & 0 & 0 \\ 0 & \tilde{l} & 0 & 0 & 0 & \tilde{m} & 0 & 0 \\ 0 & 0 & 0 & 0 & 0 & 0 & \tilde{f} & 0 \end{pmatrix}, \tag{5.9}$$

where the $\tilde{m} = 1/m$, $\tilde{M} = 1/M$, $\tilde{f} = 1/\bar{f}$ with mean duration of familiarity check $\bar{f}$, and $\tilde{lc} = 1/\bar{lc}$ with mean duration of lexical completion $\bar{lc}$. Using

$$U^+ = \begin{bmatrix} \\ U_{nn'} \\ \\ 1 \;\ldots\; 1 \end{bmatrix} \qquad \text{with} \qquad U^+ \cdot \bar{p}_n = \begin{bmatrix} 0 \\ \vdots \\ 0 \\ 1 \end{bmatrix}, \tag{5.10}$$

we can compute the non-trivial stationary probability distribution of the master equation for the E-Z Reader model under the constraints given above, Eq. (5.8). While in the original publication [29] the simulation of the E-Z Reader model was not based on the minimal process method, the results in this section demonstrate the advantage of the application of an exact simulation procedure.

5.6 The SWIFT Model

The theoretical motivation for the development of the SWIFT model was to investigate the hypothesis that all types of saccades during reading could be generated by the same process, which is the dynamical evolution of an activation field [8, 10]. The general approach to study movement preparation in a spatially-extended dynamical model has been developed by Erlhagen and Schöner [11] using partial differential equations. Their approach was based on an earlier framework on dynamics of neural fields [1]. In the case of reading, we could use a simplified mathematical description, since space is effective one-dimensional (the line of text) and discrete (the word in a given line of text).

In the SWIFT model, each word is represented by an activation variable that is tracking the current processing state. This set of lexical activations (Fig. 5.10) is the basis for saccade target selection (*where* pathway). As time evolves, the relative activations change, so that a continuous-time representation of the next saccade target exists. Whenever a saccade is prepared, the lexical activations provide a flexible mechanism for target selection. According to an influential proposal by Findlay and Walker [13] the spatial and temporal pathways of saccade generation are independent. The SWIFT model is compatible with this view, in the sense that control of fixation duration and saccade target selection are basically independent, however, interactions are due to the coupling of both pathways via the set of lexical activations.

5.6.1 Saccade Target Selection

The first key assumption in the SWIFT model is related to the selection of target words from a temporally evolving activation field. During processing, a time-dependent activation field $\{a_n(t)\}$ reflects the tendency for each word n to become selected as a saccade target. The probability $\pi_n(t)$ for target selection of word n at time t is given by the relative activation weighted by exponent γ, i.e.,

$$\pi_n(t) = \frac{(a_n(t))^\gamma}{\sum\limits_{j=1}^{N_w} (a_j(t))^\gamma}, \tag{5.11}$$

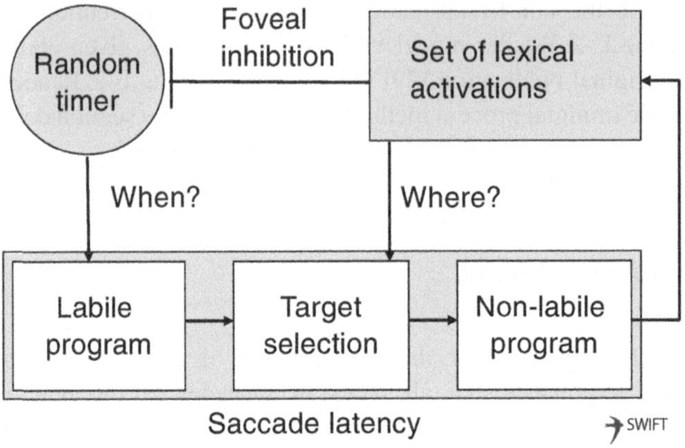

Fig. 5.10 Architecture of the SWIFT model. A temporally evolving set of activations determines the probability to select the next target word (*where* pathway). A random saccade timer with a fixed mean value is inhibited by foveal processing difficulty (*when* pathway). Saccade programming is divided into labile and non-labile stage, with target selection at the transition between the stages

which is normalized because of $\sum_{j=1}^{N_w} \pi_j(t) = 1$ for all $t > 0$. Parameter γ is important for weighting the set of lexical activations so that

$$\pi_n(t) \rightarrow \begin{cases} \text{winner-takes-all} & : \gamma \rightarrow \infty \\ \text{Luce's choice rule [20]} & : \gamma = 1 \\ \text{random selection} & : \gamma \rightarrow 0 \end{cases} . \tag{5.12}$$

For the computer implementation, we will use the *linear selection algorithm* (discussed at the end of Sect. A.2). An example for a simulated scanpath and the full time-series of lexical activation is illustrated in Fig. 5.11. As you can see in the figure, all sub-processes are implemented by discrete random walks. These processes are saccade timing, labile and non-labile saccade programming, saccade execution, and change of the set of lexical activations via lexical processing.

The time-dependent activation field required for stochastic target selection is based on parallel processing of words. While this assumption has been criticized in terms of its psychological plausibility [28], there is accumulating experimental evidence for parallel processing (e.g., [19, 36]).

5.6.2 *Numerical Simulation of SWIFT*

For the stochastic simulation of parallel processes of word processing and saccade programming, we assume parallel continuous-time random walks with Markov property in the SWIFT model. The state of the model at time t is uniquely defined

Table 5.1 List of possible stochastic transitions between internal states of the SWIFT model

	Transition from $n = (n_1, n_2, \ldots) \mapsto n' = (n'_1, n'_2, \ldots)$	
Process	Transition	Transition rate[a]
Saccade timer	$n_1 + 1$	$w_1 = r_1 \times \text{inhibition}$
Labile program	$n_2 + 1$	$w_2 = r_2, r_2 \cdot R$ (refix.), $r_2 \cdot M$ (misloc.)
Non-labile program	$n_3 + 1$	$w_3 = r_3$
Saccade execution	$n_4 + 1$	$w_4 = r_4$
Perceptual delays	$n_5 + 1$	$w_5 = r_5$
Word processing	$n_{5+j} + 1$	$w_6 = 1/\alpha \cdot \Lambda_j(t)$ (word j; $j = 1, \ldots, N_w$)

[a] Base rates: $r_1 = N_1/t_{sac}, r_2 = N_2/\tau_l, r_3 = N_3/\tau_{nl}, r_4 = N_4/\tau_{ex}, r_5 = N_5/\tau_{pc}$

by the vector $n = (n_1, n_2, \ldots, n_{4+N_w})$, where the components n_j represent the states of the sub-processes (Table 5.1).

The model is a Markov process that can be described by the general master equation,

$$\frac{\partial}{\partial t} p(n, t|n'') = \sum_{n'} \left[W_{nn'} p(n', t|n'') - W_{n'n} p(n, t|n'') \right], \tag{5.13}$$

which is determined by the transition probabilities W_{nm} for transitions $m \mapsto n$. It is important to note that m, n are state vectors in this notation. As discussed earlier, the master equation can be interpreted as a conservation equation [14, 35], where the temporal change of probability on the left side of the equation equals the *gain* in state n that is generated by transitions from neighboring states $n' \mapsto n$ and the *loss* is generated by transitions to neighboring states $n \mapsto n'$.

In Table 5.1 we list all possible transitions for the model. For word processing, we assume that change of activation for each word $j = 1, 2, 3, \ldots, N_w$ contributes its own sub-process, i.e.,

$$a_j(t) = D_j \frac{n_{5+j}}{N_{5+j}}, \tag{5.14}$$

where D_j is the word's processing difficulty, Eq. (5.20).

In this framework, temporally changing transition rates are not explicitly time-dependent (see Table 5.1), e.g., for the saccade timer. All temporal changes of transition rates are mediated by state changes of the model. For numerical simulation, we use the *minimal process method* by Gillespie (1976) [15, 33] (see Sect. A.2.3 on page 135). In Fig. 5.11 we present an example of a simulation of the SWIFT model.

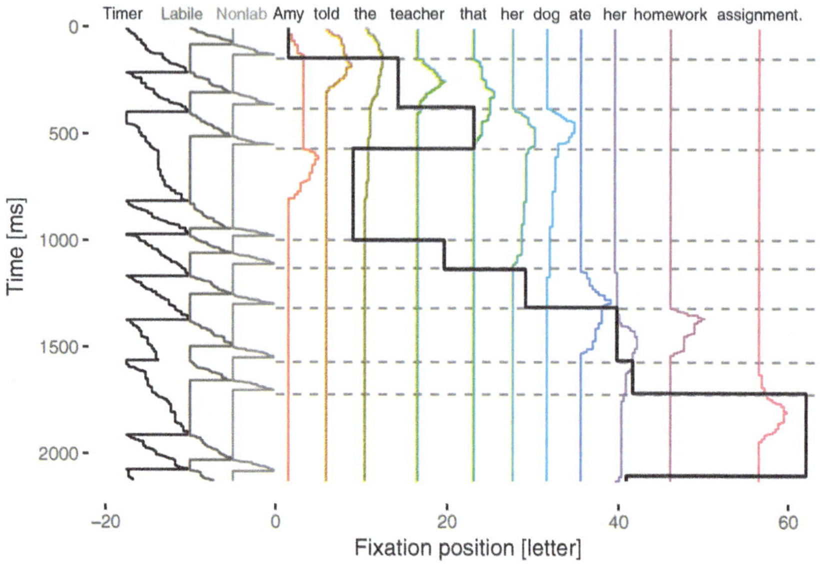

Fig. 5.11 Simulation example for the SWIFT model. The activation field (colored lines) determine the target selection probability $\pi_n(t)$. The resulting scanpath (fixation sequence) is indicated by the black line. Random walks generate saccade timer intervals and labile and non-labile saccade latencies

5.6.3 Temporal Control of Saccades and Foveal Inhibition

Processing time needed for word recognition depends on the word's frequency. It can be argued that *gaze duration*, defined as the sum of the durations of the first fixation on a word and all direct refixations, is the best measure of cognitive processing time [25]. Resulting mean gaze durations vary roughly between 200 ms and 300 ms. Since word recognition is the basis for text understanding, an adaptive mechanism for the modulation of fixation duration by word difficulty is an essential requirement for eye-movement control in reading. The necessary fixation duration for word recognition can be acquired by two opposing mechanisms: The current fixation can be prolonged by inhibited the next saccade or, alternatively, the word can be refixated to increase gaze duration.

Experimentally, it turns out that the dependence of the mean first-fixation duration on word frequency is only weak. In contrast, we find strong effects of word frequency on the probability for refixation (see Fig. 5.3). Thus, experimental data suggest that refixation is the preferred strategy to prolong gaze duration compared to saccade-inhibiting processes that extend the duration of the first fixation.

Motivated by this observation, the second central assumption in the SWIFT model is *random timing* of fixation duration with additional *foveal inhibition* [8] that delays the start of the next saccade program to extend the current fixation. We assume that foveal inhibition modulates the transition rate r_1 for transitions between elementary steps of a random walk that implements the saccade timer (left most column in Fig. 5.11).

$$r_1 = N_1/t_{sac} \times \text{inhibition}, \tag{5.15}$$

where $N_1 = 30$ is the number of state of the timer's random walk and $t_{sac} \approx 200$ ms is the mean value of the timer. Given the activation $a_k(t)$ of the fixated word k (i.e., the word in the fovea) at time t, the transition rate is modulated by the factor

$$r_1 = \frac{N_1}{t_{sac}} \cdot \left(1 + h \cdot \frac{a_k(t)}{\alpha}\right)^{-1}, \tag{5.16}$$

where $0 < h_0$ and h_1 are free parameters, which control the strength of foveal inhibition (h_0) and parafoveal inhibition (h_1) from words to the right of the current fixation, and $\lfloor . \rfloor$ returns the positive part of the argument and vanishes for negative arguments. Parameter α is the theoretical maximum value of the word's activation $a_k(t)$, so that $0 \leq a_k(t)/\alpha \leq 1$.

Using numerical simulations of the model, it can be shown that for $h > 0$, foveal inhibition can produce a modulation of the fixation duration that is in good agreement with experimental data. For difficult words (i.e., low-frequency or long words), a refixation will be generated with high probability to obtain more processing time.

5.6.4 Character-Based Visual Processing

Word recognition starts with visual processing of letters, which is done in parallel for all the letters of a given word. We define the spatial region where word activations can be influenced in the model as the *processing span*. Within this region, parallel processing is limited by the fact that processing rate depends on the letter's *eccentricity* (i.e., the distance of the letter position from the position of the current fixation). Mathematically, we define an inverted parabolic processing span from the fovea to position $-\delta_L$ on the left and to position $+\delta_R$ on the right of fixation, i.e.,

$$\lambda(\epsilon) = \lambda_0 \cdot \begin{cases} 0, & \text{for} \quad \epsilon < -\delta_L \\ 1 - \epsilon^2/\delta_L^2, & \text{for} \ -\delta_L \leq \epsilon < 0 \\ 1 - \epsilon^2/\delta_R^2, & \text{for} \ \ 0 \leq \epsilon \leq \delta_R \\ 0, & \text{for} \quad \delta_R < \epsilon \end{cases}, \tag{5.17}$$

where λ_0 is a constant given as

$$\lambda_0 = \frac{3}{2} \cdot \frac{1}{\delta_L + \delta_R}, \tag{5.18}$$

which is important to normalize the total amount of processing to one, i.e., $\int_{-\infty}^{+\infty} \lambda(\epsilon)d\epsilon = 1$.

Experimentally, a strong asymmetry of the *perceptual span* with an extension of 4 to 5 letters to the left of the fixation position and up to 15 letters to the right was found [26]. In reading experiments, however, only a mean extension of the span was estimated so far, since dynamical effects are more difficult to measure empirically.

5.6.5 Word-Based Processing Rate

Given Eqs. (5.17)–(5.18), processing rates for letters depend on spatial eccentricities. The letter j of word$_n$ is processed with rate $\lambda(\epsilon_{nj})$, if it is located at a spatial position with eccentricity $\epsilon_{nj}(t)$ relative to gaze position at time t. This letter-based processing rate must be related to the effective word-based processing rate $\lambda_n(t)$ of word$_n$ at time t.

Because of parallel processing of the letters of a given word, each letter contributes to word recognition. In the case of long words, some letters will have large eccentricities, so that their processing rate will be small according to Eq. (5.17). To capture these opposing effects in a parametric model, we make the assumption that the word-based processing rate has the form

$$\Lambda_n(t) = L_n^{-\eta} \sum_{j=1}^{L_n} \lambda(\epsilon_{nj}(t)), \tag{5.19}$$

where L_n is the length (i.e., number of letters) of word$_n$ and η is the word length exponent, with $0 < \eta < 1$. For $\eta = 0$ long words will have a processing advantage. For $\eta = 1$ word processing rate is the arithmetic mean of the letter-based processing rates (mean over all letters of a given word); therefore, we will observe a disadvantage for long words in the case $\eta = 1$. Thus, the word length exponent η is an additional free parameter in the model.

With the assumptions on spatial aspects of letter- and word-based processing rates, the temporal aspects of word processing need to be specified. As discussed for the motivation of the SWIFT model, a time-dependent activation field will provide probabilistic control of saccadic eye movements. Word-based activations $a_n(t)$ for the words of a given sentence are increasing during the initial stage of processing called *lexical processing*. After reaching the maximum of activation D_n, the activation starts to decrease (*post-lexical processing*). The maximum of activation is interpreted as processing difficulty, which is a logarithmic function of word frequency F_n, i.e.,

$$D_n = \alpha \left(1 - \beta \frac{\log F_n}{\log \bar{F}}\right), \quad \text{with} \quad \bar{F} = \max_n \{F_n\}, \tag{5.20}$$

where the parameter β determines the strength of the word frequency effect.

5.6.6 Additional Assumptions on Word Processing

In this paragraph, some additional principles of word processing in the SWIFT model are discussed.

Global decay of activation. Maintaining words in working memory during reading cannot be done without loss. Since word activations $\{a_n(t)\}$ represent the state-of-processing, we introduce a global decay of activation. If the processing rate of a word is smaller than the constant ω, then we assume a decay of activation with rate $\omega = 0.01\ \mathrm{s}^{-1}$.

Processing during saccades. During saccadic eye movements, lexical processing is paused because of *saccadic suppression* [21]. In model, we stop lexical processing, which we defined as the increasing part of the activation dynamics, during saccades, while post-lexical processing, i.e., the decreasing part of the activation dynamics, continues during saccades.

Integration of information across saccades. It is an unsolved problem how information obtained from lexical processing is integrated across saccades during reading. Therefore, we implemented a proportional reset of activation after each saccade. Mathematically, for all words that are in the lexical processing stage, we multiply the activation by the parameter ι with $0 < \iota \leq 1$ after each saccade.

5.6.7 Oculomotor Assumptions

In Chap. 4 we discussed a Bayesian model of control processes of saccadic landing positions within words. While this model would be the theoretically adequate description of oculomotor control during reading [7], it has not been integrated in cognitive models of eye-movement control so far. In current versions of the SWIFT model [10, 31] we implemented the descriptive approach for oculomotor control introduced by McConkie et al. [22].

Our assumptions on the temporal sequences of events during saccade programming are illustrated in Fig. 5.10. Our assumptions are motivated by the experimental findings by Becker and Jürgens [2] and were implemented earlier in the E-Z Reader model. Each saccade program starts with the labile stage. During the labile stage, the saccade gaze center is activated to prepare the next saccade [13]. If a new decision to start a labile saccade program happens during the labile stage, the earlier saccade program is canceled and replaced by the new program. It is the beginning of the non-labile stage that a point of no-return is passed, so the saccade can no longer be canceled or modified. After the non-labile stage of saccade programming, the execution of the saccade starts.

Modulation of the duration of the labile stage. The duration of the labile stage of saccade programming is reduced by factor R ($0 < R \leq 1$), if the saccade will become a refixation of the already fixated word. If the realized fixation position (the saccadic landing position) strongly deviates from the word center, so that the landing position will fall onto the neighboring word, then a *mislocated fixation* (see Sect. 4.3) will happen. In this case, the duration of the next saccade program will be reduced by factor M $(0 < M)$.

5.7 Exercises

Exercise 5.1 Develop computer code for the simulation of the model *E-Z Reader 1*. Try to find approximate values of model parameters to reproduce the data listed in the table of the original publication [29]. Note that the parameter given in the footnote of the table are based on a different simulation algorithm, however, these parameters should inform you about useful parameter ranges.

Exercise 5.2 Use the R code of the SWIFT model to investigate how scanpaths depend on selected model parameters. Determine the most important model parameters that influence fixation durations and the most important parameters for the control of skipping probability.

5.8 Further Reading

A comparison of different computational models of eye-movement control in reading was published in 2003 by Reichle et al. [30]. The E-Z Reader model developed by the same authors in 1998 is the currently most-cited mathematical model of eye movements in reading [29]. A three-state model was published 2001 [6], where semi-Markov processes with residence-time dependent internal states were used to fit distributions of fixation duration adequately.

The mathematical foundations of the E-Z Reader model and the SWIFT model are very different and might be characterized as *state-based* versus *dynamical* [3] models. In 1998, a detailed discussion of the philosophical foundations of state-based versus dynamical models has been published by van Gelder [34].

References

1. Amari, S.-I. (1977). Dynamics of pattern formation in lateral-inhibition type neural fields. *Biological Cybernetics, 27*(2), 77–87. https://doi.org/10.1007/BF00337259
2. Becker, W., & Jürgens, R. (1979). An analysis of the saccadic system by means of double step stimuli. *Vision Research, 19*(9), 967–983. https://doi.org/10.1016/0042-6989(79)90222-0
3. Beer, R. D. (2000). Dynamical approaches to cognitive science. *Trends in Cognitive Sciences, 4*(3), 91–99. https://doi.org/10.1016/S1364-6613(99)01440-0
4. Diacu, F., & Holmes, P. (1996). *Celestial encounters: The origins of chaos and stability*. Princeton: Princeton University Press. http://press.princeton.edu/titles/5913.html

5. Drieghe, D., Rayner, K., & Pollatsek, A. (2008). Mislocated fixations can account for parafoveal-on-foveal effects in eye movements during reading. *The Quarterly Journal of Experimental Psychology, 61*(8), 1239–1249. https://doi.org/10.1080/17470210701467953
6. Engbert, R., & Kliegl, R. (2001). Mathematical models of eye movements in reading: A possible role for autonomous saccades. *Biological Cybernetics, 85*(2), 77–87. https://doi.org/10.1007/PL00008001
7. Engbert, R., & Krügel, A. (2010). Readers use Bayesian estimation for eye movement control. *Psychological Science, 21*(3), 366–371. https://doi.org/10.1177/0956797610362060
8. Engbert, R., Longtin, A., & Kliegl, R. (2002). A dynamical model of saccade generation in reading based on spatially distributed lexical processing. *Vision Research, 42*(5), 621–636. https://doi.org/10.1016/S0042-6989(01)00301-7
9. Engbert, R., & Nuthmann, A. (2008). Self-consistent estimation of mislocated fixations during reading. *PLoS One, 3*(2), e1534, 1–6. https://doi.org/10.1371/journal.pone.0001534
10. Engbert, R., Nuthmann, A., Richter, E. M., & Kliegl, R. (2005). SWIFT: A dynamical model of saccade generation during reading. *Psychological Review, 112*(4), 777–813. https://doi.org/10.1037/0033-295X.112.4.777
11. Erlhagen, W., & Schöner, G. (2002). Dynamic field theory of movement preparation. *Psychological Review, 109*(3), 545–572. https://doi.org/10.1037/0033-295X.109.3.545
12. Findlay, J. M., & Gilchrist, I. D. (2003). *Active vision: The psychology of looking and seeing*. Oxford: Oxford University Press. https://global.oup.com/academic/product/active-vision-9780198524793
13. Findlay, J. M., & Walker, R. (1999). A model of saccade generation based on parallel processing and competitive inhibition. *Behavioral and Brain Sciences, 22*(04), 661–674. https://doi.org/10.1017/S0140525X99002150
14. Gardiner, C. (1985). *Handbook of stochastic processes*. Berlin: Springer-Verlag. http://www.springer.com/de/book/9783540707127
15. Gillespie, D. T. (1976). A general method for numerically simulating the stochastic time evolution of coupled chemical reactions. *Journal of Computational Physics, 22*(4), 403–434. https://doi.org/10.1016/0021-9991(76)90041-3
16. Kliegl, R., & Engbert, R. (2005). Fixation durations before word skipping in reading. *Psychonomic Bulletin & Review, 12*(1), 132–138. https://doi.org/10.3758/BF03196358
17. Kliegl, R., Grabner, E., Rolfs, M., & Engbert, R. (2004). Length, frequency, and predictability effects of words on eye movements in reading. *European Journal of Cognitive Psychology, 16*(1–2), 262–284. http://dx.doi.org/10.1080/09541440340000213
18. Kliegl, R., Nuthmann, A., & Engbert, R. (2006). Tracking the mind during reading: The influence of past, present, and future words on fixation durations. *Journal of Experimental Psychology: General, 135*(1), 12–35. http://dx.doi.org/10.1037/0096-3445.135.1.12
19. Kliegl, R., Risse, S., & Laubrock, J. (2007). Preview benefit and parafoveal-on-foveal effects from word n+ 2. *Journal of Experimental Psychology: Human Perception and Performance, 33*(5), 1250–1255. http://dx.doi.org/10.1037/0096-1523.33.5.1250
20. Luce, R. D. (1959). *Individual choice behavior: A theoretical analysis*. Mineola, NY: Dover Publications.
21. Matin, E. (1974). Saccadic suppression: a review and an analysis. *Psychological Bulletin, 81*(12), 899–917. https://doi.org/10.1037/h0037368
22. McConkie, G. W., Kerr, P. W., Reddix, M. D., & Zola, D. (1988). Eye movement control during reading: I. The location of initial eye fixations on words. *Vision Research, 28*(10), 1107–1118. https://doi.org/10.1016/0042-6989(88)90137-X
23. Morrison, R. E. (1984). Manipulation of stimulus onset delay in reading: evidence for parallel programming of saccades. *Journal of Experimental Psychology: Human Perception and Performance, 10*(5), 667–682. http://dx.doi.org/10.1037/0096-1523.10.5.667
24. Nuthmann, A., Engbert, R., & Kliegl, R. (2005). Mislocated fixations during reading and the inverted optimal viewing position effect. *Vision Research, 45*(17), 2201–2217. https://doi.org/10.1016/j.visres.2005.02.014

25. Rayner, K. (1998). Eye movements in reading and information processing: 20 years of research. *Psychological Bulletin, 124*(3), 372–422. http://dx.doi.org/10.1037/0033-2909.124.3.372
26. Rayner, K., Well, A. D., & Pollatsek, A. (1980). Asymmetry of the effective visual field in reading. *Perception & Psychophysics, 27*(6), 537–544. https://doi.org/10.3758/BF03198682
27. Reichle, E. D. (2015). Computational models of reading: A primer. *Language and Linguistics Compass, 9*(7), 271–284. https://doi.org/10.1111/lnc3.12144
28. Reichle, E. D., Liversedge, S. P., Pollatsek, A., & Rayner, K. (2009). Encoding multiple words simultaneously in reading is implausible. *Trends in Cognitive Sciences, 13*(3), 115–119. https://doi.org/10.1016/j.tics.2008.12.002
29. Reichle, E. D., Pollatsek, A., Fisher, D. L., & Rayner, K. (1998). Toward a model of eye movement control in reading. *Psychological Review, 105*(1), 125–157. http://dx.doi.org/10.1037/0033-295X.105.1.125
30. Reichle, E. D., Rayner, K., & Pollatsek, A. (2003). The E-Z Reader model of eye-movement control in reading: Comparisons to other models. *Behavioral and Brain Sciences, 26*(04), 445–476. http://dx.doi.org/10.1017/S0140525X03000104
31. Schad, D. J., & Engbert, R. (2012). The zoom lens of attention: Simulating shuffled versus normal text reading using the swift model. *Visual Cognition, 20*(4–5), 391–421. https://doi.org/10.1080/13506285.2012.670143
32. Schilling, H. E., Rayner, K., & Chumbley, J. I. (1998). Comparing naming, lexical decision, and eye fixation times: Word frequency effects and individual differences. *Memory & Cognition, 26*(6), 1270–1281. https://doi.org/10.3758/BF03201199
33. Schnakenberg, J. (1995). Algorithmen in der Quantenmechanik und Statistischen Physik. *Ulmen: Zimmermann-Neufang.*
34. Van Gelder, T. (1998). The dynamical hypothesis in cognitive science. *Behavioral and Brain Sciences, 21*(05), 615–628. http://dx.doi.org/10.1017/S0140525X98001733
35. Van Kampen, N. G. (1992). *Stochastic processes in physics and chemistry* (Vol. 1). Amsterdam: Elsevier. https://www.elsevier.com/books/book-series/north-holland-personal-library
36. Yan, M., Risse, S., Zhou, X., & Kliegl, R. (2012). Preview fixation duration modulates identical and semantic preview benefit in Chinese reading. *Reading and Writing, 25*(5), 1093–1111. https://doi.org/10.1007/s11145-010-9274-7

Chapter 6
Scene Viewing and Spatial Statistics

To explore an entire visual scene, we have to generate saccades several times per second to shift our gaze to interesting scene regions [6], since high-acuity vision is restricted to the fovea. Therefore, a complete theory of visual cognition must be based on understanding of ongoing visual and cognitive processes for the control of saccades in real time [7]. In scene perception, eye movements are less constrained by the structure of the visual environment than during reading, where the sequential order of words along a line of text favors scanning of words in serial order.

As a result of research on visual attention models over the past 25 years, a number of diverse computational models (termed *saliency* models) has been proposed for the prediction of fixation locations in an input image [10, 11, 19, 20]. From a biological point of view, such static saliency maps are implausible, since the full detail of the maps could only be available for eye guidance *after* exploration of the scene, however, the saliency maps are computed before scene exploration as a basis for eye guidance. Thus, a more realistic combined model of visual processing and eye-movement control would involve a fixation-position dependent saliency map, which (i) initially contains only low-spatial frequency information about the overall gist of the scene and which (ii) is modified and enriched with more detailed information (e.g., about specific objects) during each new fixation.

Moreover, the sequence of saccades during scene viewing shows complicated statistical behavior. Instead of sampling statistically independent fixation positions from a saliency map, models include an additional inhibitory mechanism to prevent the eye-movement control system from continually refixating the point of highest saliency [9]. A fundamental biological process for this type of *inhibitory-of-return* has been found earlier in simple visual search paradigms [12, 13]. However, a direct manifestation of inhibition-of-return could not be reproduced in natural scene-viewing task [17, 18].

In this situation of inconclusive experimental evidence, mathematical models are needed to investigate the interaction between attentional control and inhibitory tagging for eye-movement control during scene viewing. In a recent paper, a

R. Engbert, *Dynamical Models In Neurocognitive Psychology*,
https://doi.org/10.1007/978-3-030-67299-7_6

dynamical model for the control of saccades has been developed with the research strategy (i) to reproduce most of the rich statistical patterns from eye-movement experiments and (ii) to investigate the importance of inhibition-of-return for gaze guidance in natural situations [5]. The chapter starts with an introduction to spatial statistics and eye-movement control and continues with dynamical modeling of scanpath-generating processes.

6.1 Gaze Pattern During Scene Viewing

It is a basic property of the human visual system that our ability to perceive visual detail falls off rapidly from the center of the visual field (i.e., the current gaze position) into a low-resolution visual surround. Therefore, we reorient the fovea about three to four times per second. Information processing can only be done during fixations (i.e., periods of relative gaze stability with ongoing fixational eye movements) because of *saccadic suppression* during the saccades [14]. Gaze control is the process of generating the scanpath over time under perceptual, cognitive, and behavioral constraints. Gaze patterns are experimental data with rich statistical properties. Since the main focus has been on fixation positions during the last decades, theoretical models have been developed to predict the two-dimensional (2D) density of fixations [3].

The estimation of the 2D density from experimentally observed fixation positions is typically done via kernel density estimation. The procedure is illustrated in Fig. 6.1 for one-dimensional data sampled from a normal distribution. In the kernel density estimation approach a smooth kernel function is centered at each data point to overcome the discreteness of the data. By summing over the kernel functions centered at actual observations, we obtain a density estimate. The basic kernel estimator can be written as

$$\hat{f}_{\text{KDE}}(x) = \frac{1}{N}\sum_{i=1}^{N} K\left(\frac{x - x_i}{h}\right) , \tag{6.1}$$

where $K(x)$ is the smoothing kernel (e.g., a Gaussian function) and h is the bandwidth parameter. In the following, we will use 2D kernel density estimates for the estimation of the 2D density of fixations (gaze positions) observed in eye-tracking experiments.

An example for a typical scanpath generated by a human observer is given in Fig. 6.2a, where fixations are connected by the red lines to highlight the sequence (scan path). Each fixation contributes to the local probability density (computed via kernel density estimation) and color-coded in Fig. 6.2b.

Additionally, fixation durations are characterized by durations in addition to positions. In such *marked point processes*, we study the statistical dependence between locations and additional properties *marks*) of the point process. In the

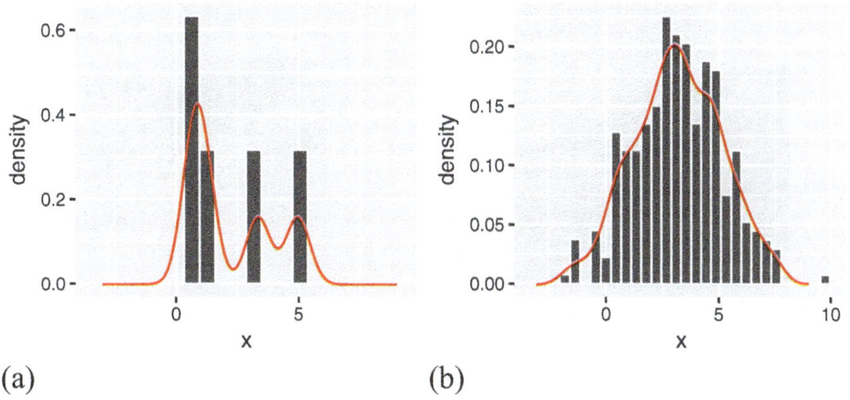

Fig. 6.1 Illustration of kernel density estimation for a Gaussian smoothing kernel. (**a**) Estimate produced for $N = 5$ samples from a normal distribution. (**b**) Resulting estimate for $N = 300$. In both examples, a bandwidth of $h = 0.5$ was used

next section, we will discuss point processes and their basic application to eye-movement data. Using the framework of spatial point processes, we will be able to investigate statistical dependencies between realization of the point processes, in particular, spatial correlation functions.

6.2 Point Process Statistics

Point processes are stochastic models for the generation of irregular point patterns. While the term "process" is often related to a time-dependent phenomenon, point processes are, in most cases, used for the description of time-independent spatial patterns. Mathematically, a point process N is a *random set* of points $x_1, x_2, \ldots$, i.e.,

$$N = \{x_i\} \qquad \text{or} \qquad N = \{x_1, x_2, \ldots\}\,, \tag{6.2}$$

where the set can be finite or infinite. At the same time, the symbol N is used as a *counting measure*. For a subset B of $\mathbb{R}^d$ (with $d = 2$ for the two-dimensional case), $N(B)$ is the number of points in B. It is assumed that the counting measure holds the fundamental property of additivity,

$$N(B_1 \cup B_2) = N(B_1) + N(B_2) \tag{6.3}$$

for disjoint subsets B_1 and B_2. The value $N(B)$ for a set B is a random variable, where we consider multiple realizations of the point process under consideration. As in classical statistics, we consider the mean $\mathbf{E}(N(B))$, where $\mathbf{E}$ is the symbol for

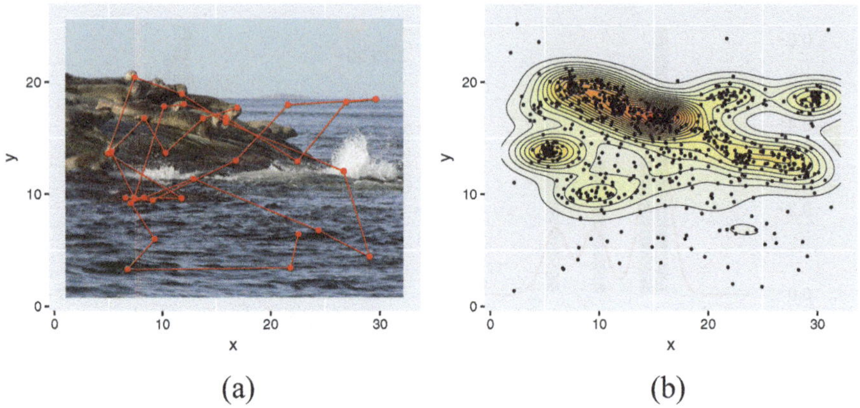

Fig. 6.2 Typical gaze pattern and representation as spatial point process. (**a**) Series of fixations generated in a single trial by a human observer (lines indicate the scan path). (**b**) Same sequence shown as realization of a marked point process, where the two-dimensional density of fixations is coded by color

expectation. Therefore,

$$\mathbf{E}(N(B)) = \text{mean number of points of } N \text{ in } B\ . \tag{6.4}$$

Based on this concept, we introduce the *intensity measure*

$$\Lambda(B) = \mathbf{E}(N(B))\ , \tag{6.5}$$

which is a (deterministic) function operating on sets (i.e., a measure). For all practical purposes, we can make a continuity assumption, so that a density called *intensity function* $\lambda(x)$ exists with

$$\Lambda(B) = \int_B \lambda(x)\mathrm{d}x\ . \tag{6.6}$$

The assumption of continuity is invalid, for example, if the points are arranged on a regular grid.

Following Illian et al. [8], a point process N is called *stationary* or *homogeneous* if N and the translated point process N_x have the same distribution for all translations x. For $N = \{x_1, x_2, \ldots\}$, the translated process is given by $N_x = \{x_1 + x, x_2 + x, \ldots\}$. The property of having the same distribution refers to the statement

$$\mathbf{P}(N(B_1) = n_1, \ldots, N(B_k) = n_k) = \mathbf{P}(N_x(B_1) = n_1, \ldots, N_x(B_k) = n_k) \tag{6.7}$$

for k arbitrary subsets B_1, $B_2, \ldots$

A simple, but important model for spatial point processes is the *homogeneous Poisson point process* with density λ. For this process, (i) the number of points in any region B follows the Poisson distribution with mean $\lambda v(B)$ (where $v(.)$ denotes the area/volume of B) and (ii) the numbers of points of N in k disjoint sets form k independent random variables.

To describe situations with spatial inhomogeneity, the homogeneous Poisson process can be generalized, when the constant intensity λ is replaced by a position-dependent intensity $\lambda(x)$. In this case of the *inhomogeneous Poisson process*, the number of points of N in any bounded set B has a Poisson distribution with mean $\int_B \lambda(x)\mathrm{d}x$.

Now, we apply spatial point processes to scene-viewing data. In Fig. 6.3a, we show experimental data from a human observer during exploration of an image. In the first step, we must estimate the 2D density (intensity) of the gaze patterns from all participants for the image. This is typically done by kernel density estimation; for the data in the figure, we determined the bandwidth parameter σ according to Scott's rule [16] and obtained a value of $\sigma \approx 2°$ (in Fig. 6.3d, we used $\sigma = 5°$ to illustrate the 2D density by grey level). The 2D density is an important foundation to run simulations of the null models, i.e., the inhomogeneous and homogeneous Poisson point processes.

A simulation of a corresponding inhomogeneous Poisson process with the same density as the original data is plotted in Fig. 6.3b. First-order statistics of the point patterns in Fig. 6.3a and b are the same. In the case of the homogeneous Poisson process simulation (Fig. 6.3c), the density is constant over the observation window. Local density variations indicated by the grey shading in Fig. 6.3c are due to random fluctuations.

6.3 Second-Order Statistics of Gaze Patterns

The homogeneous and inhomogeneous Poisson point processes are attractive as simple models that do not include spatial interactions. Therefore, these models are important to generate data corresponding to certain null hypotheses, when spatial interactions in real data are investigated. Such interactions can be analyzed using second-order statistics.

Second-order statistics are based on the pair density $\rho(x_1, x_2)$, which gives the probability $\rho(x_1, x_2)\mathrm{d}x_1\mathrm{d}x_2$ for finding points in each of two disks b_1 and b_2 with diameters $\mathrm{d}x_1$ and $\mathrm{d}x_2$, respectively. For the characterization of point patterns, it turns out that the pair density is typically a function of the distance $r = \|x_1 - x_2\|$. As a consequence, the pair density $\rho(x_1, x_2) = \rho(r)$ is a continuous function of r. An estimation of the pair density is obtained by kernel density computation, i.e.,

$$\hat{\rho}(r) = \sum_{x_1, x_2 \in W}^{\neq} \frac{k(\|x_1 - x_2\| - r)}{2\pi r A_{\|x_1 - x_2\|}}, \tag{6.8}$$

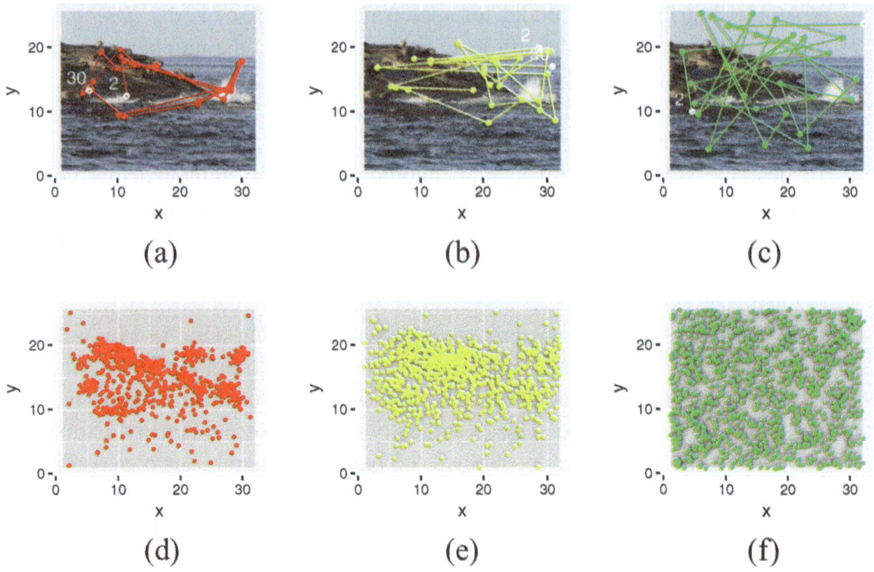

Fig. 6.3 Point process patterns for experimental data, inhomogeneous and homogeneous point processes. (**a**) Experimental data (red dots) from a single trial. (**b**) Simulated data from inhomogeneous point process. (**c**) Simulated data from homogeneous point process. (**d**) Full experimental dataset of all fixations on the image. (**e**) Simulated inhomogeneous point process from density $\lambda(x)$, computed via kernel density estimation with bandwidth $\sigma = 5°$. (**f**) Simulated data from homogeneous point process

where $k(.)$ is a kernel function and $A_{\|x_1 - x_2\|}$ is an edge correction at distance $\|x_1 - x_2\|$. In the following, we use the Epanechnikov kernel [8], i.e.,

$$k(x) = \begin{cases} \frac{3}{4h}(1 - \frac{x^2}{h^2}), & \text{for } -h \leq x \leq h \\ 0, & \text{otherwise} \end{cases}, \tag{6.9}$$

for numerical computations.

From the pair density, the *pair correlation function* $g(r)$ can be derived by normalization of the pair density with respect to first-order intensity $\hat{\lambda}$. Thus, the estimator for the pair correlation is obtained by $\hat{g}(r) = \rho(r)/\hat{\lambda}^2$. The pair correlation function $g(r)$ can be interpreted in the following way: If two disks with areas dA_1 and dA_2 are considered at distance r, then the probability to find a point simultaneously in each disk is given by

$$g(r)\lambda^2 dV_1 dV_2 \,. \tag{6.10}$$

The interpretation of the numerical values computed by the pair correlation function for a given point pattern is straightforward: For a random pattern with complete spatial randomness, the pair correlation function will yield $\hat{g}(r) \approx 1$

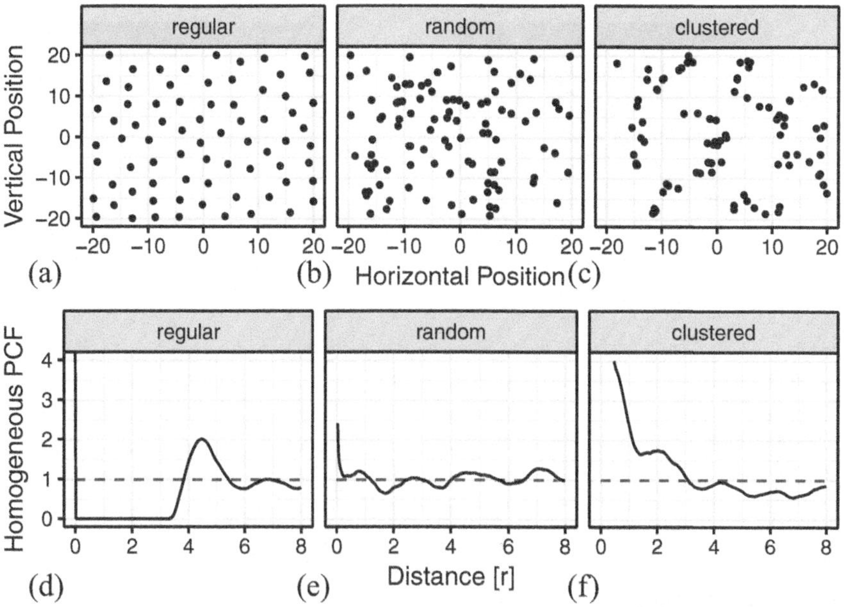

Fig. 6.4 Illustration of a homogeneous point process with different types of spatial correlations. **(a,b,c)** Examples of regular, random, and aggregated (clustered) point patterns. **(d,e,f)** Corresponding pair correlation functions

across a broad range of distances r. If $\hat{g}(r) > 1$, then pairs of fixations are more abundant than on average at distance r, which indicates spatial aggregation at distance r. If $\hat{g}(r) < 1$, then pairs of fixations are less abundant than on average at a distance r compared to complete spatial randomness. As a result, the pair correlation function $\hat{g}(r)$ quantifies how the selection of a particular point location is influenced by realizations at other locations at distance r. This statistical framework turns out to be a powerful tool for the analysis of gaze patterns (Fig. 6.4).

So far, we introduced the pair correlation function $g(r)$, which was limited to the study of homogeneous point processes. However, the concept can be generalized. Using the inhomogeneous pair correlation function $g_{inh}(r)$, we can remove the first-order inhomogeneity from the second-order spatial statistics by the definition of the estimator

$$\hat{g}_{inh}(r) = \sum_{x_1,x_2 \in W}^{\neq} \frac{1}{\hat{\lambda}(x_1)\hat{\lambda}(x_2)} \frac{k(\|x_1 - x_2\| - r)}{2\pi r A_{\|x_1 - x_2\|}} . \tag{6.11}$$

Since the inhomogeneous 2D density $\hat{\lambda}(x)$ is needed in this equation, the estimation of $\hat{g}_{inh}(r)$ involves two steps: First, the overall intensity $\hat{\lambda}(x)$ must be estimated from all observations obtained for a given scene. Second, the pair correlation function of a single trial is computed with respect to the inhomogeneous density

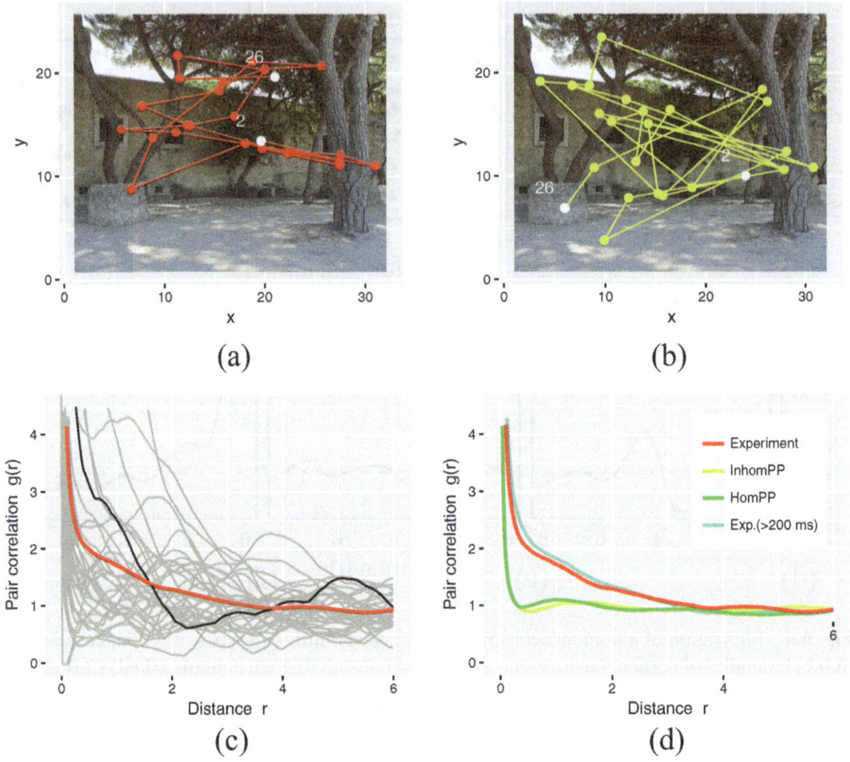

Fig. 6.5 Pair correlation function (PCF) for gaze patterns. (**a**) Single-trial data from experiments. (**b**) Realization of a corresponding inhomogeneous Poisson process. (**c**) Single-trial PCFs (grey line) and mean PCF (red). The trial shown in (**a**) is highlighted (black). (**d**) Mean PCFs for experimental data, inhomogeneous and homogeneous Poisson processes, and experimental data without short fixation (< 200 ms)

of the full data set. In this procedure, we borrow strength from the full dataset of all observations to be able to compute reliable estimates of the inhomogeneity.

For the statistical analysis of spatial correlations, the inhomogeneous and homogeneous Poisson point processes are an important basis (Fig. 6.5a,b). Estimated pair correlation functions (PCF) for single trials show considerable variance because of the limited observation length of 10 to 20 fixations in each trial (Fig. 6.5c). The mean PCF, computed as average over different observers for a given image, turns out to be a smooth function of the distance r. In the next step of our analysis, we compare the mean PCF between experimental data and the Poisson null models (Fig. 6.5d). As a result, we observe significant spatial aggregation at short length scales ($r \lesssim 4°$) for the experimental data, while this effect is clearly absent in the two null models. Note, however, that the PCFs diverge for $r \to 0$, which is an artifact of the estimator and cannot be interpreted as the presence of positive correlations (aggregation) at small length scales.

The result for the inhomogeneous Poisson process indicates that the inhomogeneity of the underlying 2D density did not create the deviations from $\hat{g}(r) = 1$. However, could it be possible that the inhomogeneity was overestimated by the density estimation of $\hat{\lambda}(x)$? This is tested by running the same analyses on the simulated data from a homogeneous Poisson process. In the case of overestimation of the inhomogeneity, we would have obtained a value $\hat{g}(r) < 1$, which is not the case. Thus, the statistical null models support our conclusion that the experimental data show spatial aggregation at short length scales.

For a statistical test of this effect of spatial aggregation, a typical measure [8], the *PCF deviation*, is defined as

$$\Delta_g = \int_0^\infty (\hat{g}(r) - 1)^2 \mathrm{d}r \ , \tag{6.12}$$

which quantifies the mean deviation of the pair correlation from the uncorrelated case with $g(r) = 1$. Because of the divergence of the estimator, Eq. (6.11), for $r \to 0$, the integral in Eq. (6.12) should be evaluated between finite boundaries, e.g., $0.1° < r < 5°$. A more detailed discussion of a related statistical analysis is published in the original literature [5].

The complex statistical patterns underlying gaze positions in scene viewing motivated the development of a dynamical model discussed at the end of the chapter. Before we explain this model, we need to introduce ordinary differential equations as a mathematical model for time-dependent processes, which will be done in Appendix B.

6.4 Dynamic Modeling of Scene Viewing

The pair correlation function is a theoretically well-grounded measure of the sequential dependencies in gaze patterns. Because of these dependencies, the estimation of the probability distribution for the next fixation will be based on all previous fixations in a given trial. In this section we describe the SceneWalk model that computes such probability distributions.

The SceneWalk model is composed of two independent processing streams of saccade planning that are related to attention allocation (excitatory pathway) [9, 10] and inhibition-of-return (inhibitory pathway) [12, 13], respectively (Fig. 6.6). It is important to note that the SceneWalk model does not include a saliency model. We start with an empirically determined distribution of gaze positions (empirical saliency map). One could replace this distribution by a model-generated saliency map, of course.

The excitatory pathway starts with empirical saliency, multiplied by a Gaussian attention window around the current fixation position. This product is used as a fixation-position dependent saliency map. Such a fixation-position dependent map is important for biological plausibility, since visual acuity declines towards the

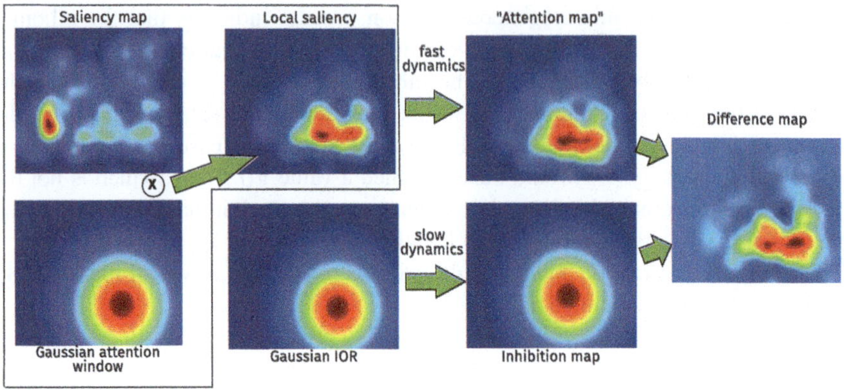

Fig. 6.6 Illustration of the two processing streams of the SceneWalk model [5]. The time-dependence of the attention pathway and the pathway for inhibitory tagging generates a time-dependent potential $u(x, t)$, from which a 2D probability distribution for the upcoming saccade target is read out. A local saliency map is used as the input for the build-up of activation in the attention map. An inhibition map is subtracted, which builds up more slowly using a constant-shape Gaussian around the current fixation as input. The difference map undergoes thresholding and normalization to produce the final distribution $u(x, t)$ for the probabilistic selection of the next saccade target

periphery. Similarly, for the inhibitory pathway, a simple Gaussian around the current fixation indicates the currently fixated region of the visual field. Local saliency and inhibitory Gaussian are both implicitly time-dependent because of saccades that shift fixation positions.

Let us denote the current fixation position by $\mathbf{x}_f = (x_f, y_f)$. First, we compute the two Gaussian distributions centered at $\mathbf{x}_f$ on a grid of size $L \times L$. For the attentional pathway we implement a Gaussian aperture G_A with standard deviation σ_A to read out the local saliency map. The inhibitory pathway is based on a Gaussian G_F with standard deviation σ_F to induce activation that drives the model away from the currently fixated area to new parts of the visual field. For a grid position (x, y) these Gaussians can be written as

$$G_{A/F}(x, y; x_f, y_f) = \frac{1}{2\pi\sigma_{A/F}^2} \exp\left(-\frac{(x - x_f)^2 + (y - y_f)^2}{2\sigma_{A/F}^2}\right), \tag{6.13}$$

where A indicates the (excitatory) attention map and F denotes the fixation map (inhibitory tagging).

As the next step, we specify the change over time for both the attention map $A(t) = \{A_{ij}(t)\}$ and the fixation map $F(t) = \{F_{ij}(t)\}$, where the indices $1 \leq \{i, j\} \leq L$ are related to grid positions across the full image. Rates of activation changes of the two maps are determined by the two parameters ω_A and ω_F, resp. Using the time-independent salience map $S = \{S_{ij}\}$ and the Gaussians G_A and G_F defined in Eq. (6.13), the temporal evolutions of both activation maps are given by

$$\frac{dA_{ij}(t)}{dt} = -\omega_A A_{ij}(t) + \omega_A \frac{S_{ij} \cdot G_A(x_i, y_j; x_f, y_f)}{\sum_{kl} S_{kl} \cdot G_A(x_k, y_l; x_f, y_f)} \tag{6.14}$$

$$\frac{dF_{ij}(t)}{dt} = -\omega_F F_{ij}(t) + \omega_F \frac{G_F(x_i, y_j; x_f, y_f)}{\sum_{kl} G_F(x_k, y_l; x_f, y_f)}, \tag{6.15}$$

where $\sum_{kl}$ indicates the sum over all grid-points (k, l).

The input $G_{A/F}$ changes because of gaze shifts $\mathbf{x}_f \mapsto \mathbf{x}'_f$; therefore the structure of the right-hand side of Eqs. (6.14) and (6.15) is simple and we can solve this system analytically between two saccadic gaze shifts. This can be seen by comparing Eqs. (6.14–6.15) to Eq. (B.8) on page 143.

For initial maps A_0 and F_0 at the start of a given fixation at time t_0, the solution of the differential equations can be written as

$$A(t) = \frac{G_A S}{\sum G_A S} + e^{-\omega_A (t-t_0)} \left(A_0 - \frac{G_A S}{\sum G_A S} \right) \tag{6.16}$$

and

$$F(t) = \frac{G_F}{\sum G_F} + e^{-\omega_F (t-t_0)} \left(F_0 - \frac{G_F}{\sum G_F} \right); \tag{6.17}$$

note that indices were omitted to simplify the notation. The solutions are given by an exponential change from the map represented at the beginning of the fixation towards the input map. Thus, Eqs. (6.16) and (6.17) can be used to iterate the activation maps for attention and inhibitory tagging from fixation to fixation using a single computational time step t.

To run a simulation, both activation maps of the model must be initialized. Therefore, we prepare the maps with a uniform distribution (normalized to one) and adjust the magnitude of the input such that the equilibrium size of the maps is automatically normalized to one as well. Using this procedure, the total activation of the attention map and of the map for inhibitory tagging is held at a constant value of one over time.

Given the two activation maps $A(x, t)$ and $F(x, t)$, we are looking for a combined map $u(x, t)$ to select the upcoming saccade target. A straightforward approach is to use the difference of the attentional and inhibitory maps after thresholding and normalization. For a flexible weighting of activation in each map, activations are raised to power λ for the attention map A and to power γ for the fixation map F, respectively. Next, each map is normalized [4]. Finally, the map for inhibitory tagging is multiplied by a factor c_F and subtracted from the attention map. As a result, we define a time-dependent potential $u_{ij}(t)$ for target selection, i.e.,

$$u_{ij}(t) = \frac{[A_{ij}(t)]^\lambda}{\sum_{kl}[A_{kl}(t)]^\lambda} - c_F \frac{[F_{ij}(t)]^\gamma}{\sum_{kl}[F_{kl}(t)]^\gamma} \,. \tag{6.18}$$

Evaluating the power-functions of activations at each grid-point changes not only the weighting between different local maxima but also shrinks or widens the individual peaks. Therefore, we re-parametrized the model parameters by choosing

$$\lambda\sigma_A'^2 = \sigma_A^2 \qquad \gamma\sigma_F'^2 = \sigma_F^2 \,, \tag{6.19}$$

which defines parameters σ_A' and σ_F' as the standard deviations of the Gaussians when mapped through the nonlinearity directly.

To transform $u_{ij}(t)$ into a probability distribution, we consider the positive part only and normalize the potential to obtain a unit integral. First, we assume a function that maps u to an intermediate u^*, which is positive everywhere, i.e.,

$$u^*(u) = \begin{cases} u & u > 0 \\ 0 & u \le 0. \end{cases} \tag{6.20}$$

Second, we add a uniform distribution using a weighting factor ζ to produce a function $\pi(i, j) > 0$ for each lattice position,

$$\pi(i, j) = (1 - \zeta)\frac{u^*_{ij}}{\sum_{kl} u^*_{kl}} + \zeta \frac{1}{\sum_{kl} 1} \,, \tag{6.21}$$

which can be used as a target selection probability. Thus, the original function u is transformed into an activation map that gives a probability ($\ge \zeta / \sum_{kl} 1$) for next fixation position (saccade target). The distribution $\pi(i, j)$ generates the probability of any grid-point to be the next fixation target, given the previous sequence of fixations.

In summary, there are equations of motion, i.e., the differential equations, Eqs. (6.14—6.15), which determine the saccade targeting probability, Eq. (6.21). This temporal evolution is motivated by the assumption of two physiologically independent processing streams of attention and inhibitory tagging. Saccade target selection is a stochastic process that determines the next fixation position as a 2D probability from the interaction of these temporally evolving activation fields.

Given the complexity of the fixation sequences, the SceneWalk model is parsimonious in terms of number of model parameters (see Table 6.1). There are two parameters for the decays of activations of the two maps and two parameters for the activation spans around the current fixation position. Finally there are different weighting parameters. The original parameter estimates were taken from the first publication of the model [5], while different numerical values were obtained for a revised model using the maximum-likelihood estimation technique.

Table 6.1 Overview on model parameters of the SceneWalk model

Parameter name	Original estimate	MLE	Description
ω_A	6.607	2.4×10^{30}	Decay attention map
ω_F	0.00903	1.9298	Decay fixation map
σ_A	4.88	5.9082	Span attention
σ_F	3.9436	4.5531	Span inhibitory tagging
γ	0.3*	44.780	Exponent attention
λ	1*	0.8115	Exponent fixation
c_F	1*	0.3637	Weighting
ζ	–	0.0722	

6.5 SceneWalk Model: Results

The SceneWalk model [5] is able to simulate scanpaths which can be analyzed using the same measures that we applied to experimental data. A qualitative comparison is shown in Fig. 6.7a,b, where a number of fixations is plotted for a given image together with a sample scanpath.

Next, we compare the distributions of generated saccade lengths (Fig. 6.7c) for experimental data and a range of models. For experimental data (red line) and the dynamical SceneWalk model (blue line), the distributions peak at relatively short saccade lengths (about 5°). This is in contrast with the homogeneous point process model which also produces a single-humped distribution, but overestimates the mean saccade length dramatically (green line). Interestingly, for the inhomogeneous point process, this effect is still present (yellow line).

The pair correlation functions (Fig. 6.7d) indicate that experimental data (red line) and SceneWalk model simulations both produce clearly aggregated scanpaths, while the statistical control models of homogeneous and inhomogeneous point processes fluctuate around $g(r) \approx 1$. The corresponding PCF deviations, Eq. (6.12), are plotted in Fig. 6.8b. For the statistical evaluation we computed the single-trial PCF deviations and produced a boxplot for all models and the experimental data. As a result, there is good agreement between SceneWalk model and experimental data, while the homogeneous and inhomogeneous point processes are not significantly different from an uncorrelated process.

But how well do the models reproduce the 2D density of fixations? To quantify this, we need a measure for the deviation between experimental and simulated densities of fixations. To compute deviations between 2D densities P_{kl} and Q_{kl} at grid position (k, l), we used the Kullback–Leibler divergence, a symmetric version of the information gain [2], i.e.,

$$\Delta_{KLD} = \frac{1}{2} \sum_{k,l} \left(P_{kl} \log \frac{P_{kl}}{Q_{kl}} + Q_{kl} \log \frac{Q_{kl}}{P_{kl}} \right) . \tag{6.22}$$

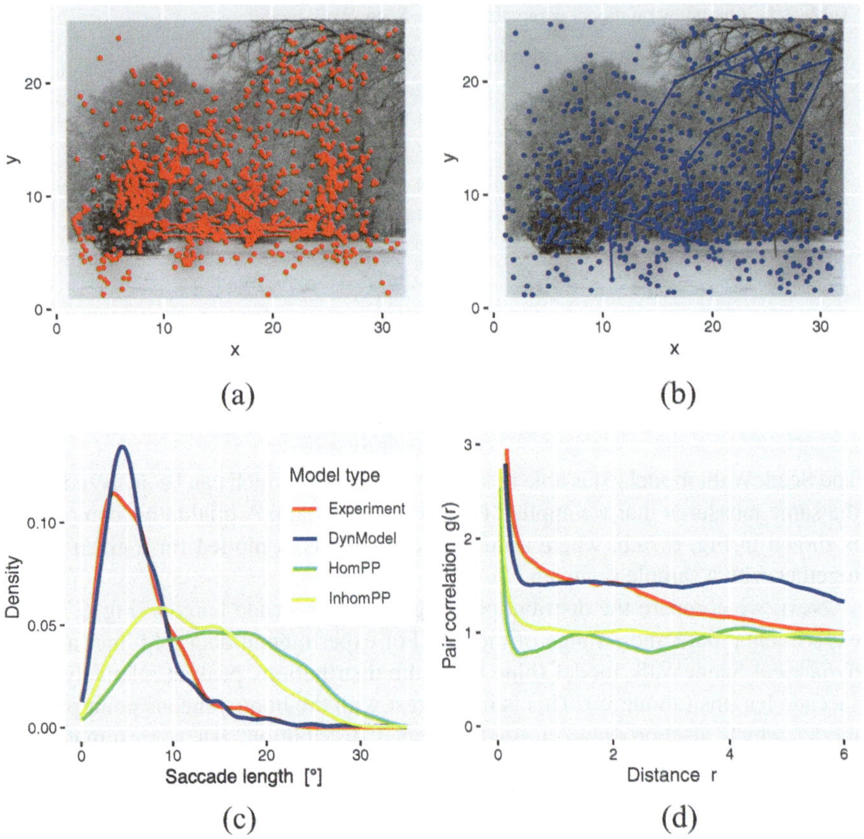

Fig. 6.7 Comparison between experimental data and SceneWalk simulations. (**a**) Experimental fixation positions (red dots) and a typical scanpath (red line). (**b**) Simulated fixation positions (blue dots) and typical scanpath. (**c**) Distributions of saccade lengths for a dataset of 30 picture and 35 observers and an equivalent number of simulations. (**d**) Corresponding mean pair correlations functions

The baseline for Δ_{KLD} can be obtained from experimental data by splitting the full dataset into two halves. Taking the first half of data as the density P_{kl} and the second as the density Q_{kl}, we can estimate how well the data can be reproduced by a different set of observers. As one can see from the corresponding boxplot in Fig. 6.8a, the SceneWalk model is as good as the inhomogeneous point process with respect to the Kullback–Leibler divergence. Thus, with the SceneWalk model, we implemented a model which reproduces a measure of first-order statistics (2D density) as well as a measure of second-order statistics (PCF) for a spatial point process.

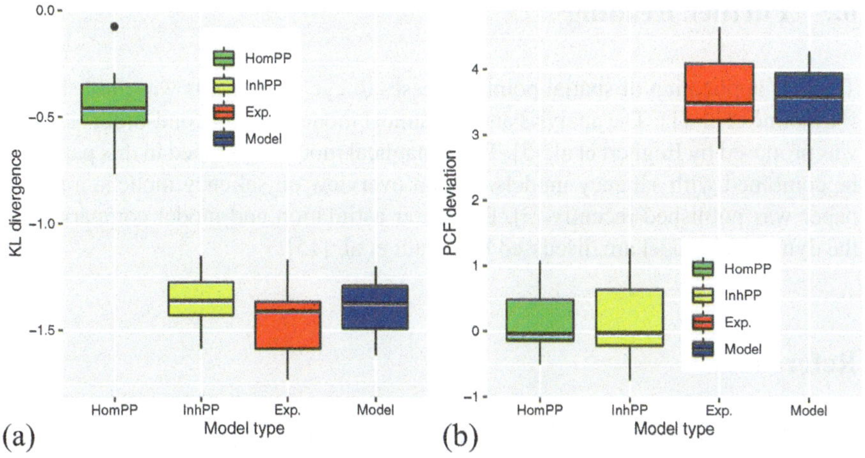

Fig. 6.8 Quantitative measures for goodness-of-fit of the SceneWalk model. (**a**) Kullback-Leibler divergence, Eq. (6.22). (**b**) PCF deviation, Eq. (6.12), from uncorrelated fixation patterns

6.6 Exercises

Exercise 6.1 (Kernel Density Estimation) Reproduce the type of plots shown in Fig. 6.1 by varying (i) the underlying sampling distribution (Gaussian, mixture of Gaussians, gamma distribution, etc.), (ii) the bandwidth of the kernel, and (iii) the number of data samples.

Exercise 6.2 (Compute Samples from Inhomogeneous and Homogeneous Point Processes) Estimate the 2D density of fixation positions recorded for a typical image. Sample the same number of data points from the density to obtain data from the corresponding inhomogeneous point process. Compare the results for a homogeneous point process (see Fig. 6.3).

Exercise 6.3 (Pair Correlation Function for Correlated Scanpaths) Use the homogeneous point process model from Exercise 6.2 and introduce sequential interactions. Sample data points sequentially and accept the next data sample with a probability that depends on the distance to the previous sample. Compute the pair correlation function.

Exercise 6.4 (SceneWalk Model Simulations) Use the R implementation of the SceneWalk model to generate a large number of scanpaths ($N = 100$). (a) Compute the distribution of saccade lengths. (b) Compare your results with the distributions of saccade lengths obtained for the inhomogeneous and homogeneous point process and for the experimental data.

6.7 Further Reading

The first application of spatial point processes to eye movements was published by Barthelmé et al. [1]. The analysis and dynamical modeling of second-order statistics was proposed by Engbert et al. [5]. The dynamical model developed in this paper can be combined with saliency models. For an overview on saliency models, a review paper was published recently [3]. Parameter estimation and model comparison of the dynamical model are discussed by Schütt et al. [15].

References

1. Barthelmé, S., Trukenbrod, H., Engbert, R., & Wichmann, F. (2013). Modeling fixation locations using spatial point processes. *Journal of Vision, 13*(12), 1, 1–34. https://doi.org/10.1167/13.12.1.
2. Beck, C., & Schögl, F. (1993). *Thermodynamics of chaotic systems: An introduction.* Cambridge Nonlinear Science Series (vol. 4). Cambridge: Cambridge University Press. https://www.cambridge.org/gb/academic/subjects/physics/nonlinear-science-and-fluid-dynamics/thermodynamics-chaotic-systems-introduction.
3. Borji, A., & Itti, L. (2013). State-of-the-art in visual attention modeling. *IEEE Transactions on Pattern Analysis and Machine Intelligence, 35*(1), 185–207. https://doi.org/10.1109/TPAMI.2012.89.
4. Carandini, M., & Heeger, D. J. (2012). Normalization as a canonical neural computation. *Nature Reviews Neuroscience, 13*(1), 51–62. https://doi.org/10.1038/nrn3136.
5. Engbert, R., Trukenbrod, H. A., Barthelmé, S., & Wichmann, F. A. (2015). Spatial statistics and attentional dynamics in scene viewing. *Journal of Vision, 15*(1), 14, 1–17. https://doi.org/10.1167/15.1.14.
6. Findlay, J. M., & Gilchrist, I. D. (2003). *Active vision: The psychology of looking and seeing.* Oxford: Oxford University Press. https://global.oup.com/academic/product/active-vision-9780198524793.
7. Henderson, J. M. (2003). Human gaze control during real-world scene perception. *Trends in Cognitive Sciences, 7*(11), 498–504. https://doi.org/10.1016/j.tics.2003.09.006.
8. Illian, J., Penttinen, A., Stoyan, H., & Stoyan, D. (2008). *Statistical analysis and modelling of spatial point patterns* (vol. 70). John Wiley & Sons. https://eu.wiley.com/WileyCDA/WileyTitle/productCd-0470014911.html.
9. Itti, L., & Koch, C. (2001). Computational modelling of visual attention. *Nature Reviews Neuroscience, 2*(3), 194–203. https://doi.org/10.1038/35058500.
10. Itti, L., Koch, C., & Niebur, E. (1998). A model of saliency-based visual attention for rapid scene analysis. *IEEE Transactions on pattern analysis and machine intelligence, 20*(11), 1254–1259. https://doi.org/10.1109/34.730558.
11. Kienzle, W., Franz, M. O., Schölkopf, B., & Wichmann, F. A. (2009). Center-surround patterns emerge as optimal predictors for human saccade targets. *Journal of Vision, 9*(5), 7, 1–15. https://doi.org/10.1167/9.5.7.
12. Klein, R. M. (2000). Inhibition of return. *Trends in Cognitive Sciences, 4*(4), 138–147. https://doi.org/10.1016/S1364-6613(00)01452-2.
13. Klein, R. M., & MacInnes, W. J. (1999). Inhibition of return is a foraging facilitator in visual search. *Psychological Science, 10*(4), 346–352. https://doi.org/10.1111/1467-9280.00166.
14. Matin, E. (1974). Saccadic suppression: a review and an analysis. *Psychological Bulletin, 81*(12), 899–917. https://doi.org/10.1037/h0037368.

15. Schütt, H. H., Rothkegel, L., Trukenbrod, H. A., Reich, S., Wichmann, F. A., & Engbert, R. (2017). Likelihood-based parameter estimation and comparison of dynamical cognitive models. *Psychological Review, 124*(5), 505–524. https://doi.org/10.1037/rev0000068.
16. Scott, D. W. (1992). *Multivariate density estimation: theory, practice, and visualization* (vol. 383). John Wiley & Sons. https://eu.wiley.com/WileyCDA/WileyTitle/productCd-0471697559,subjectCd-ST70.html.
17. Smith, T. J., & Henderson, J. M. (2009). Facilitation of return during scene viewing. *Visual Cognition, 17*(6-7), 1083–1108. https://doi.org/10.1080/13506280802678557.
18. Smith, T. J., & Henderson, J. M. (2011). Does oculomotor inhibition of return influence fixation probability during scene search? *Attention, Perception, & Psychophysics, 73*(8), 2384–2398. https://doi.org/10.3758/s13414-011-0191-x.
19. Torralba, A., Oliva, A., Castelhano, M. S., & Henderson, J. M. (2006). Contextual guidance of eye movements and attention in real-world scenes: the role of global features in object search. *Psychological Review, 113*(4), 766–786. https://doi.org/10.1037/0033-295X.113.4.766.
20. Tsotsos, J. K., Culhane, S. M., Kei Wai, W. Y., Lai, Y., Davis, N., & Nuflo, F. (1995). Modeling visual attention via selective tuning. *Artificial Intelligence, 78*(1), 507–545. https://doi.org/10.1016/0004-3702(95)00025-9.

Chapter 7
Bimanual Coordination and Coupled Oscillations

The coordination of limbs during execution of human motor behavior can be studied in experimental paradigms on simple rhythmic movements. While there is a bewildering complexity of coordinated movements in human motor control [9], the presence of a number of elementary coordination rules can be demonstrated [11]. Despite their simplicity, simple rhythmic movements are important to everyday activity, for example, during walking, swimming, or cycling. In this chapter, we focus in bimanual movements, in particular, movements with synchronized and alternating movement sequences.

Rhythmic (or oscillatory) behavior is ubiquitous in biological systems, from physiological rhythms and the synchronization of fireflies to human movement patterns [2, 10]. For the mathematical description of biological systems, the theory of nonlinear oscillations plays a key role [4]. In experimental psychology, a paradigmatic study of bimanual coordination [6] has been explained theoretically by coupled nonlinear oscillators [3], which we will discuss in detail in this chapter.

For model development, we will have to discuss ordinary differential equations for the mathematical description of coupled, nonlinear oscillations. Since analytical solutions are often very difficult (and practically impossible) to obtain, numerical solutions are most important for studying nonlinear ordinary differential equations. Some of the basic concepts in analytical and geometrical techniques, however, are very helpful for understanding the qualitative behavior of the solutions.

7.1 An Experiment on Bimanual Coordination

In his, now classical, experiment on rhythmic bimanual coordination of the two index fingers, Scott Kelso [6] observed a spontaneous transition between two dynamical modes that can be characterized by their anatomical symmetry.

R. Engbert, *Dynamical Models In Neurocognitive Psychology*,
https://doi.org/10.1007/978-3-030-67299-7_7

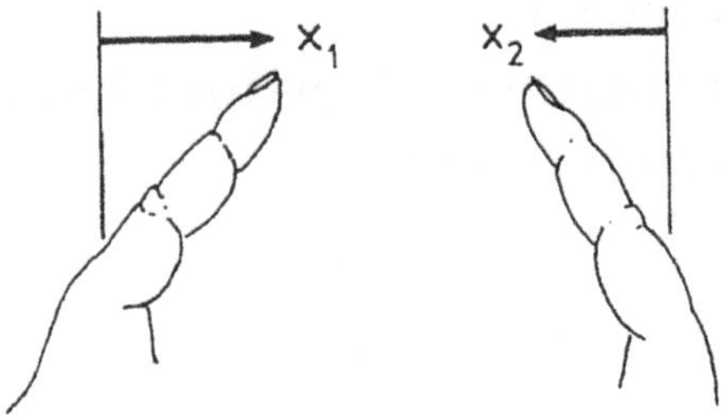

Fig. 7.1 Definition of coordinates for bimanual movements. The symmetrical mode refers to the activation of homologous muscle groups; it is the only stable movement pattern, which can be maintained by all participants at high frequency (from: [3])

Participants were instructed to start rhythmic finger movements in an anti-symmetrical[1]motion with low frequency. Now, the participants were required to gradually increase the frequency of the movements. The participants' finger movements shifted suddenly to a symmetrical movements patterns (Fig. 7.1).

As a starting for the mathematical analysis of the transition between the two qualitatively different movement patterns, we have to find key dynamical variables, which characterize the dynamics. Let us consider a simple periodic process that can be described by a cosine function with constant amplitude A,

$$x(t) = A\cos(\omega t) = A\cos(\phi(t)) , \tag{7.1}$$

where the phase of the oscillation $\phi(t) = \omega t$ is a linearly increasing function of time t. It is important to note that we can add multiples of 2π to a phase variable without any consequences. The slope ω of this increasing phase is the frequency of the oscillation. The relation between the period T and the frequency ω is $\omega = 2\pi/T$. If we aim at a parsimonious description of the relation of two oscillators $x_1(t)$ and $x_2(t)$, the *relative phase* (or phase difference) turns out to be a useful and interesting dynamical variable. Therefore, we assume that there are two cosine functions with constant amplitudes A_1 and A_2, Eq. (7.1), for the two oscillators ($x_1(t)$ left finger, $x_2(t)$ right finger),

$$x_1(t) = A_1\cos(\omega t + \phi_1) \tag{7.2}$$

$$x_2(t) = A_2\cos(\omega t + \phi_2) , \tag{7.3}$$

where the constants ϕ_1 and ϕ_2 (the so-called *phase-shifts*) are the phases of the oscillators at time $t = 0$. Since the frequencies are identical, these two oscillations show a constant relative phase ϕ over time,

$$\phi = \phi_2 - \phi_1 . \tag{7.4}$$

[1]The term anti-symmetrical refers to the fact that non-homologous (or non-similar) muscles are activated in the left and right hands.

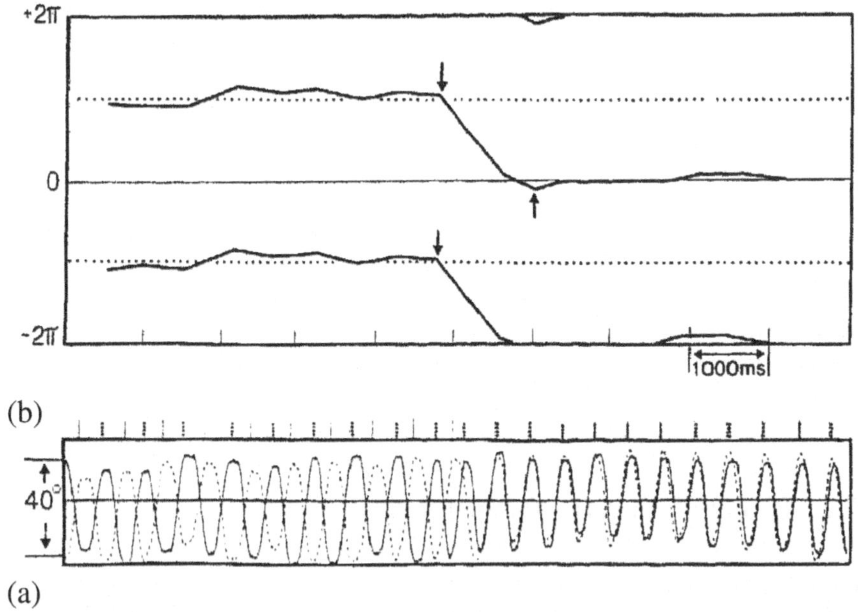

Fig. 7.2 Typical data of the bimanual coordination experiment by Kelso [6]. (**a**) The phase of the two fingers oscillate and, suddenly, switch from the anti-symmetrical to the symmetrical mode of coordination. (**b**) The relative phase ϕ can be used to characterize the transition between the two dynamical modes (from: [3])

In a convenient coordinate system (Fig. 7.1), the relative phase is $\phi = 0$ in the *symmetrical* mode, which is also called an *in-phase* movement, and $\phi = \pi$ in the *anti-symmetrical* pattern, which is correspondingly referred to as an *out-of-phase* motion pattern. The transition from the anti-symmetrical to the symmetrical mode can identified by a transition from $\phi = \pi$ to $\phi = 0$ (Fig. 7.2). Before we start with the development of a mathematical model for the dynamics of the relative phase ϕ, we briefly summarize the main experimental findings:

1. There are two stable states (attractors) for low movement frequency. Participants can perform the anti-symmetrical ($\phi = \pm\pi$) as well as the symmetrical ($\phi = 0$) mode.
2. When movement frequency is gradually increased, a spontaneous transition from the anti-symmetrical to the symmetrical mode is observed. The anti-symmetrical movement can no longer be maintained.
3. If movement frequency is reduced after the transition to the symmetrical mode, no transition back to the anti-symmetrical mode is found.
4. If participants start in the symmetrical movement pattern, no transition is observed.

Therefore, the relative phase ϕ is an important variable that characterizes the mode of the movement task. In the first step of model development, we will focus on the dynamical states of the relative phase. Based on the resulting model, the transition dynamics can be explained. Furthermore, this model for the relative phase will be an important starting point for a coupled oscillator model, which describes the oscillating finger movements explicitly.

In the next sections, we will discuss the concept of potentials for 1D movements and develop a potential-based model for the relative phase (Sects. 7.2 and 7.3). Next, we will introduce and study numerical solutions for ordinary differential equations (Sect. B.2), harmonic oscillations (Sect. B.3), and nonlinear oscillations (Sect. B.4). Finally, we will discuss the Haken-Kelso-Bunz model [3], a model of nonlinear coupled oscillators for the transition from anti-symmetrical to symmetrical finger movements (Sect. 7.4).

7.2 Potentials and 1D Vector Fields

For some one-dimensional dynamical systems, e.g., $\dot{x} = x - x^3$, the concept of the *potential function* can be helpful in studying the dynamics of the systems. Potential functions are motivated by mechanical problems in physics. Let us consider the general case of a 1D ordinary differential equation,

$$\dot{x} = f(x) . \tag{7.5}$$

If we think of the dynamics described by Eq. (7.5) as a particle sliding down a hill in a landscape, the topography of the landscape is given by the potential. Mathematically, the negative derivative (or gradient) of the potential V plays the role of the mechanical force,

$$f(x) = -\frac{\partial V}{\partial x} , \tag{7.6}$$

which acts on the particle moving in the potential $V(x)$ at position x.

Let us consider the example of $f(x) = x - x^3$. Integrating the right-hand side $f(x)$ of the equation give us the potential,

$$V(x) = -\frac{1}{2}x^2 + \frac{1}{4}x^4 + C , \tag{7.7}$$

where C is an arbitrary constant. The relation between the function $f(x)$ and the potential $V(x)$ is illustrated in Fig. 7.3.

Equilibrium points (also called steady states) can be characterized by extrema of the potential V, i.e., $dV/dx = 0$, which are points of vanishing force f at

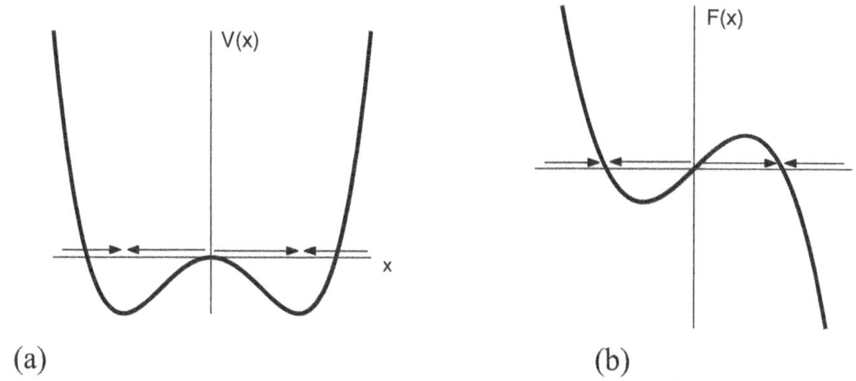

Fig. 7.3 Potential and vector field. (**a**) Graph of the potential and equilibrium points given as local minima. (**b**) The force of the potential, calculated as negative derivative, where the equilibrium points are obtained as intersection of the function with the x-axis

intersections with the horizontal axis, i.e., $f(x) = 0$. Thus, in an equilibrium point, there is no force acting on the particle.

In Fig. 7.3, the forces resulting from Eq. (7.6) are drawn as arrows on the x-axis. These arrows represent a one-dimensional vector field. At each point of the x-axis, the arrow symbolizes the driving force a particle would be exposed to, if placed in the potential.

Equilibrium points are very important in the analysis of movements. In our example of Kelso's [6] bimanual finger movements, the mode of symmetrical and anti-symmetrical movements are equilibrium states of the coordination dynamics. In the next section, we will discuss a possible potential for the mathematical description of the dynamics of relative phase.

7.3 The HKB Potential

In the first step of their development of a mathematical model of the transition in bimanual finger movements, Haken et al. [3] use a potential function (Sect. 7.2) to describe the experimental findings on relative phase ϕ in terms of equilibrium points. Therefore, we are searching for a potential $V(\phi)$ with

$$\dot{\phi} = -\frac{\partial V}{\partial \phi} \,. \tag{7.8}$$

The relative phase ϕ is *invariant* under addition of multiples of 2π. Since we are seeking a potential function to describe the dynamics of the relative phase, the potential must show this invariance as well, i.e.,

$$V(\phi + 2\pi k) = V(\phi) \,, \tag{7.9}$$

where $k = 0, \pm 1, \pm 2, \ldots$ Furthermore, exchanging the role of the hands does not change the results, which implies a symmetrical potential, characterized by the property

$$V(\phi) = V(-\phi) \, . \tag{7.10}$$

With these two boundary conditions, we can construct a potential function $V(\phi)$ as a sum of two cosine functions,

$$V(\phi) = -a\cos(\phi) - b\cos(2\phi) \, , \tag{7.11}$$

where a and b are free parameters. The resulting form of the potential depends on the ratio b/a, which is illustrated in the panels of Fig. 7.4.

Now we calculate the minima of the potential function $V(\phi)$ as a function of the parameters a and b. A necessary condition for existence of a minimum is that the slope of the potential function is zero, i.e., the derivative of $V(\phi)$ with respect to ϕ vanishes,

$$\frac{\partial V}{\partial \phi} = a \sin \phi + 2b \sin 2\phi = 0 \, . \tag{7.12}$$

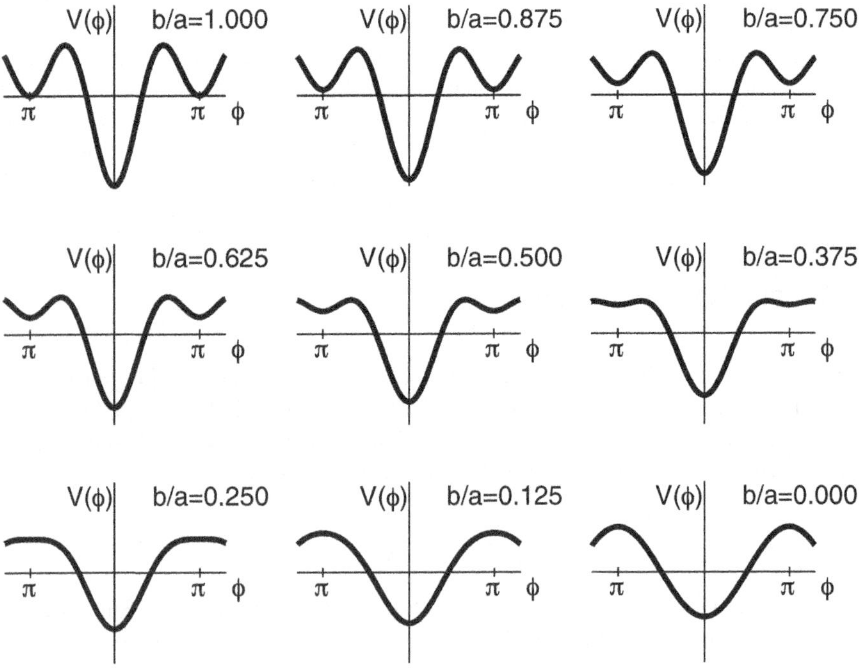

Fig. 7.4 The potential $V(\phi)$, Eq. (7.11), for varying values of the ratio b/a

Next we rearrange Eq. (7.12) using a theorem from trigonometry,

$$\sin 2\phi = 2\sin\phi\cos\phi \tag{7.13}$$

and obtain $\sin\phi\,(a + 4b\,\cos\,\phi) = 0$. This equation holds, if at least one of the two factors is zero, i.e.,

$$\sin\,\phi = 0 \qquad \text{or} \qquad a + 4b\,\cos\,\phi = 0\,. \tag{7.14}$$

The condition is fulfilled for $\phi = 0$ or $\phi = \pm\pi$, which are the extremal points of the potential. The maxima of the potential are given by the second equation, which we can write as $\cos\phi = -a/(4b)$. If these maxima disappear, we observe the transition of the anti-symmetrical mode, $\phi = \pm\pi$, to the symmetrical mode, $\phi = 0$. Since the values of the cosine function are bounded by -1 and 1, the maxima disappear, if

$$\left|\frac{a}{4b}\right| > 1\,. \tag{7.15}$$

Thus, we were able to derive a criterion for the transition between the two modes of oscillation in the potential model.

At this point of analysis, it is unclear how the parameters a and b (or even the ratio a/b) relate to the experiment. We will see in the next section that the condition in Eq. (7.15) will be useful for the derivation of a coupled oscillator model of Kelso's experiment.

7.4 *The HKB Nonlinear Oscillator Model

In the previous section, we studied the qualitative transition from anti-symmetric to symmetric finger movements in Kelso's experiment. However, instead of developing an explicit model of the movements of both fingers, we introduced the concept of relative phase. Using this minimal mathematical model, Haken, Kelso, and Bunz (1985) were able to formulate a potential function for the relative phase, which could be used to describe the stability properties of the finger movements. In this section, we aim at a more detailed mathematical model of Kelso's experiment in terms of coupled, nonlinear oscillators, which directly describes the movement trajectories of both hands, $x_1(t)$ and $x_2(t)$.

In the previous sections, we discussed the necessary mathematical techniques of numerically solving differential equations, which occur in the modeling of oscillatory behavior. In order to describe human finger movements, we need to use a nonlinear oscillator to obtain *self-sustained* oscillations. This can be shown experimentally in perturbation studies: After small perturbations [5, 12], human finger movements return to their earlier amplitude and frequency.

As an example for such a nonlinear, self-sustained oscillator, we studied the van-der-Pol equations in Sect. B.4. However, in the van-der-Pol oscillator, the movement amplitude does not dependent on the frequency ω. Since we observe in experiments that the amplitude decreases with increasing movement frequency, Haken et al. [3] use a different oscillator in their original paper, given by the *Rayleigh* equations,

$$\ddot{x} + \epsilon(\dot{x}^2 - \omega_0^2 r_0^2)\dot{x} + \omega^2 x = 0 \,. \tag{7.16}$$

In the Rayleigh oscillator, the friction term is velocity dependent. Therefore, the velocity $\dot{x}$ is stabilized to the value $\omega_0^2 r_0^2$. A consequence of this mechanism is that the amplitude of the oscillations decreases for increasing frequency. Because of the relation between frequency ω and movement amplitude r, the transition can be characterized by a critical amplitude r_c, where the transition is observed.

To describe our problem of bimanual coordination, we use a system of two coupled, self-sustained oscillators of the Rayleigh type,

$$\ddot{x}_1 + \epsilon(\dot{x}_1^2 - \omega_0^2 r_0^2)\dot{x}_1 + \omega^2 x_1 = I_{12}(x_1, x_2, \dot{x}_1, \dot{x}_2) \tag{7.17}$$

$$\ddot{x}_2 + \epsilon(\dot{x}_2^2 - \omega_0^2 r_0^2)\dot{x}_2 + \omega^2 x_2 = I_{21}(x_1, x_2, \dot{x}_1, \dot{x}_2) \,, \tag{7.18}$$

where the coupling terms are symmetrical, i.e., $I_{12} = -I_{21}$, and generally depend on positions $x_{1,2}$ and velocities $\dot{x}_{1,2}$ of the two fingers. In the simplest case, we implement a linear coupling of the form

$$I_{12} = \alpha(x_1 - x_2) \tag{7.19}$$

$$I_{21} = \alpha(x_2 - x_1) = -I_{12} \,. \tag{7.20}$$

Following Haken et al. [3], it turns out that a linear coupling term cannot reproduce the correct phase relations. If we use two coupled Rayleigh oscillators, a nonlinear coupling is necessary to reproduce the experimentally observed results. Furthermore, the coupling term depends on both positions $x_{1,2}(t)$ and velocities $\dot{x}_{1,2}(t)$. A nonlinear coupling of the form

$$I_{12} = (\dot{x}_1 - \dot{x}_2)[\alpha + \beta(x_1 - x_2)^2] \tag{7.21}$$

$$I_{21} = (\dot{x}_2 - \dot{x}_1)[\alpha + \beta(x_2 - x_1)^2] = -I_{12} \tag{7.22}$$

leads—after some calculations which involve several approximations—to an equation for the relative phase ϕ,

$$\dot{\phi} = (\alpha + 2\beta r^2)\,\sin\phi - \beta r^2\,\sin 2\phi \,, \tag{7.23}$$

where the parameter r is the amplitude of the oscillation. Equation (7.23) is of the required type,

$$\dot{\phi} = -a \sin \phi - 2b \sin 2\phi \tag{7.24}$$

that we obtained previously in our investigation of the potential function, Eq. (7.11). If we compare these two equations, we find that $a = \alpha + 2\beta r^2$ and $b = -2\beta r^2$. Putting together these relations with the condition for the transition in the potential function, Eq. (7.15), gives us an equation for the critical amplitude r_c of the oscillation,

$$\frac{a}{4b} = \frac{\alpha + 2\beta r_c^2}{2\beta r_c^2} = 1 \tag{7.25}$$

$$\rightarrow \quad r_c = \sqrt{\frac{\alpha}{4\beta}} \, . \tag{7.26}$$

These calculations are based on a combination of the potential function approach for modeling stability of relative phase and general properties of coupled nonlinear oscillators.

For the numerical study of the HKB coupled oscillators, we write the two equations of second order, Eqs. (7.17), as an equation system four differential equations of first order (see Sect. B.3). The resulting system of equation is given as

$$\dot{x}_1 = x_3 \tag{7.27}$$

$$\dot{x}_2 = x_4 \tag{7.28}$$

$$\dot{x}_3 = -\epsilon(x_3^2 - \omega_0^2 r_0^2)x_3 + (x_3 - x_4)[\alpha + \beta(x_1 - x_2)^2)] \tag{7.29}$$

$$\dot{x}_4 = -\epsilon(x_4^2 - \omega_0^2 r_0^2)x_4 + (x_3 - x_4)[\alpha + \beta(x_1 - x_2)^2)] \, . \tag{7.30}$$

To study the transition from anti-symmetric to symmetric finger movements, we introduce a linearly increasing frequency of the oscillations (from 1.17 to 3.05 s^{-1}). we also use slightly asymmetric initial conditions ($x_1(0) = 0.95$, $x_2(0) = -1$, and $\dot{x}_1(0) = \dot{x}_2(0) = 0$) to enforce the transition. A trajectory for this system of equations is shown in Fig. 7.5. Parameters and initial values for the simulation are summarized in the figure caption.

In summary, qualitative transitions in rhythmic finger movements can be described successfully a system of nonlinear coupled oscillators. The first step of model development by Haken et al. [3] was to study a potential for a qualitative description of the movement's qualitative states, the anti-symmetric and symmetric states and their stability properties. This model constrained the possible choices of coupling terms in the coupled oscillator model. Finally, the numerical simulation demonstrated the viability of the approach.

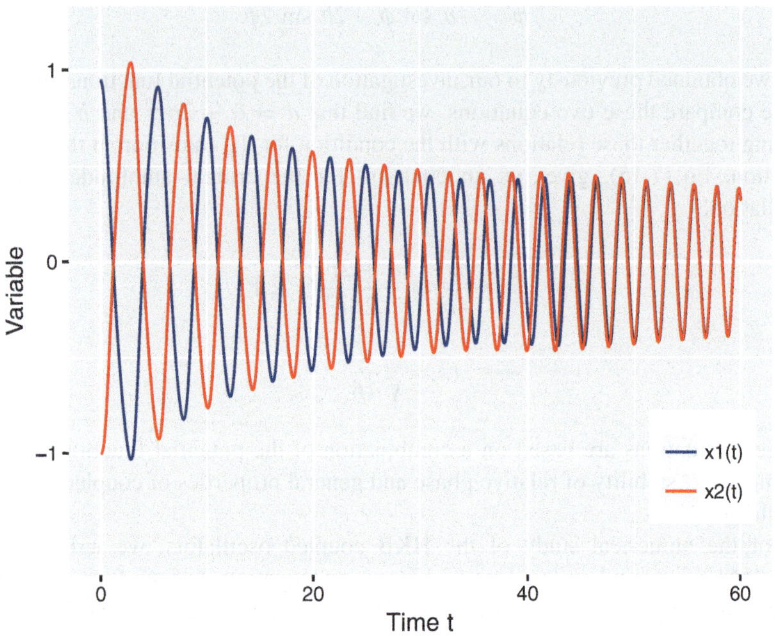

Fig. 7.5 A numerical simulation of the phase transition in voluntary cyclical hand movements. The parameters of the system of coupled oscillators, Eqs. (7.17) and (7.21), were fixed at $\epsilon = 1$, $\omega_o^2 r_0^2 = 1$, $\alpha = -0.2$ and $\beta = 0.2$. The frequency ω of the oscillations changes from 1.17 to 3.05. To induce the transition, the initial conditions of the oscillators were chosen with a small deviation from the anti-symmetrical mode, $x_1(0) = 0.95$, $x_2(0) = -1$ and $\dot{x}_1(0) = \dot{x}_2(0) = 0$

7.5 Exercises

Exercise 7.1 (Numerical Solution of the Lorenz Equations) Nonlinear ODEs can have attractors of very complicated geometrical structure. Study numerical solutions of the Lorenz equations [8],

$$\begin{aligned} \dot{x} &= \sigma(y - x) \\ \dot{y} &= \rho x - y - xz \,, \\ \dot{z} &= -\beta z + xy \end{aligned} \tag{7.31}$$

with parameters $\sigma = 10$, $\beta = 8/3$, and $\rho = 28$. Use the fourth-order Runge-Kutta method to compute a trajectory for a time $T = 50$ using $\Delta t = 0.01$ with initial condition $(x_0, y_0, z_0) = (0, 10, 10)$. Plot the resulting solution as a 3D trajectory $(x(t), y(t), z(t))$ using the R package `scatterplot3d`.

Exercise 7.2 (Sensitive Dependence on Initial Conditions) This is a continuation of the previous exercise. Choose a second initial condition with $z_0' = z_0 + 10^{-9}$

and solve the system again. (a) Visualize the x-component of the trajectories for both initial conditions in a single plot over time t. (b) Plot the difference of the x-components. Describe your results.

Exercise 7.3 (Numerical Study of Coupled Oscillators in the HKB Model) Implement the numerical solution of the system of nonlinear coupled oscillators in the HKB model and reproduce the transition shown in Fig. 7.5. Parameter values and initial conditions are given in the figure caption.

7.6 Further Reading

A different experimental approach to bimanual coordination is rhythm production, in particular, the generation of polyrhythms [7], where transitions between several movement modes were observed and a mathematical model was proposed [1].

References

1. Engbert, R., Scheffczyk, C., Krampe, R. T., Rosenblum, M., Kurths, J., & Kliegl, R. (1997). Tempo-induced transitions in polyrhythmic hand movements. *Physical Review E, 56*(5), 5823–5833. https://doi.org/10.1103/PhysRevE.56.5823
2. Glass, L., & Mackey, M. C. (1988). *From clocks to chaos: the rhythms of life*. Princeton: Princeton University Press.
3. Haken, H., Kelso, J. S., & Bunz, H. (1985). A theoretical model of phase transitions in human hand movements. *Biological Cybernetics, 51*(5), 347–356. https://doi.org/10.1007/BF00336922
4. Kaplan, D., & Glass, L. (2012). *Understanding nonlinear dynamics*. New York: Springer. https://link.springer.com/book/10.1007/978-1-4612-0823-5
5. Kelso, J. (1981). On the oscillatory basis of movement. *Bulletin of the Psychonomic Society, 18*, 63.
6. Kelso, J. (1984). Phase transitions and critical behavior in human bimanual coordination. *American Journal of Physiology-Regulatory, Integrative and Comparative Physiology, 246*(6), R1000–R1004. http://ajpregu.physiology.org/content/246/6/R1000
7. Krampe, R. T., Kliegl, R., Mayr, U., Engbert, R., & Vorberg, D. (2000). The fast and the slow of skilled bimanual rhythm production: parallel versus integrated timing. *Journal of Experimental Psychology: Human Perception and Performance, 26*(1), 206–233. https://doi.org/10.1037/0096-1523.26.1.206
8. Lorenz, E. N. (1963). Deterministic nonperiodic flow. *Journal of the Atmospheric Sciences, 20*(2), 130–141. https://doi.org/10.1175/1520-0469(1963)020<0130:DNF>2.0.CO;2
9. Rosenbaum, D. A. (2009). *Human Motor Control*. Academic Press. http://psycnet.apa.org/psycinfo/2010-06119-000
10. Strogatz, S. H., & Stewart, I. (1993). Coupled oscillators and biological synchronization. *Scientific American, 269*(6), 102–109. https://www.scientificamerican.com/article/coupled-oscillators-and-biological
11. Swinnen, S. P. (2002). Intermanual coordination: from behavioural principles to neural-network interactions. *Nature Reviews Neuroscience, 3*(5), 348–359. https://doi.org/10.1038/nrn807
12. Yamanishi, J.-I., Kawato, M., & Suzuki, R. (1980). Two coupled oscillators as a model for the coordinated finger tapping by both hands. *Biological Cybernetics, 37*(4), 219–225. https://doi.org/10.1007/BF00337040

Chapter 8
Epilog: Dynamical Models of Cognition

Looking back on the contents of this book, I would like to discuss three general aspects of mathematical modeling of cognitive processes. First, many decisions must be made, before we can actually start to build a mathematical model [3, 17]. Some of these decisions, highlighted by Neil Gershenfeld [10], will be discussed in the next section. Second, from the perspective of classification of different modeling work, we will compare the different types of models, using the examples investigated in this course. Third, there are uses and abuses of mathematical models as discussed by Robert May [18]. These problems and questions are not always easy to recognize and sometimes difficult to address adequately.

8.1 Decisions Before You Can Start Your Modeling Project

The selection of the type of mathematical model is most important, since many of the answers, which we will obtain from a model, are limited by the specific model type. For example, the Bayesian computations for initial fixation positions in reading (Chap. 4) or the spatial point processes for scene viewing (Chap. 6) are time-independent. Consequently, we will not get new insights on temporal aspects of the processes from analyses of such time-independent models. Therefore, in generally, the choice of the model's type should be motivated, crucially, by the questions we are asking.

Most mathematical models will be designed, built, or derived, after multiple levels of decisions [10] were made; some of these levels are highlighted in Fig. 8.1. It might be impossible to generate a complete and coherent list of levels. However, these meta-modeling concepts are crucial for understanding our own models and for evaluation of other researchers' models.

R. Engbert, *Dynamical Models In Neurocognitive Psychology*,
https://doi.org/10.1007/978-3-030-67299-7_8

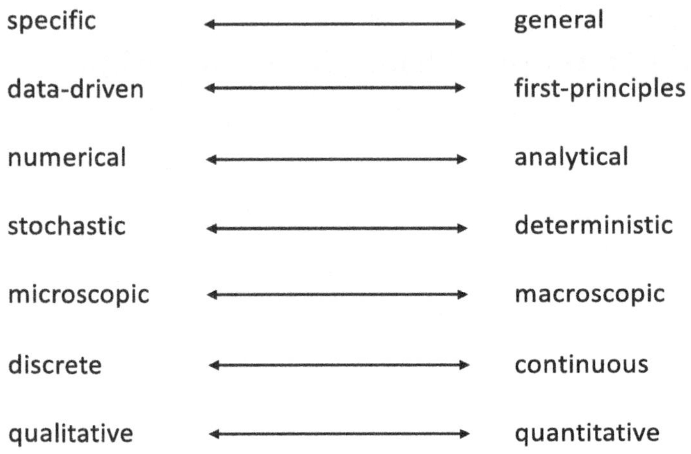

Fig. 8.1 Typical levels of description in mathematical models (after: [10])

8.1.1 Data-Driven Models vs. Models from First Principles

The development of theories often starts with data-driven models. In cognitive science, important examples are statistical models, e.g., regression models that test interactions of variables and significances of observed effects or distributional models on a certain random variable like reaction time. In contrast to these models, dynamical models typically implement assumptions (*first principles*) on variables that drive the temporal evolution of a process. These assumptions are often more fundamental than the observed variables. For example, the transitions rules of the random-walk model are more fundamental than properties (e.g., the distribution) of reaction times, since the transition rules *explain* properties of reaction times. Therefore, data-driven models and dynamical models are complementary in the development of a theory for a cognitive phenomenon.

8.1.2 Microscopic vs. Macroscopic

Models starting with fundamental processes are often *microscopic*. In the simple random-walk model, transitions between internal states can be interpreted as elementary (or microscopic) decision processes [21]. Therefore, the model links microscopic processes to the macroscopic distribution of decision times that is easily accessible from experiments. In this example, the microscopic models can also be related to neural activity, which is not necessarily possible in all microscopic models.

8.1.3 *Qualitative vs. Quantitative*

Many mathematical models will generate both quantitative and qualitative results. For example, the E-Z Reader model (Sect. 5.3) predicts fixation durations observed from experiments, so a *quantitative* measurement result can be explained by the model. However, there are also qualitative aspects of the model. The E-Z Reader model [22] predicts that fixation durations before a skipping saccade will be increased—a *qualitative* prediction. In this case, there is clear experimental evidence against this prediction [16].

8.1.4 *Numerical vs. Analytical*

Mathematical models are often implemented on a digital computer to obtain quantitative predictions using *numerical* simulations. Other models are analytical, such as the Bayesian model of within-word landing positions (Sect. 4.3), where we were able to calculate an analytical formula for the launch-site effect [5]. Analytical results are difficult or impossible to obtain for many numerical models, e.g., there is no analytical solution for the skipping probabilities in the SWIFT model (Sect. 5.6). Some models are both analytical and numerical. In the delayed random-walk model, we compared the analytical and numerical solutions for the stationary probability distribution (Sect. 2.3).

8.1.5 *Stochastic vs. Deterministic*

Random walks are classical examples of *stochastic* models [4, 15]. Each run generates a stochastic time evolution: A single simulated reaction time is not necessarily representative for the distribution, which can only be obtained from many runs. In contrast, the Haken-Kelso-Bunz model for rhythmic finger movements [12] is fully *deterministic*; the initial conditions (in this case positions, $x_{1,2}$, and velocities, $\dot{x}_{1,2}$, of both fingers at time $t = 0$) determine the future states. Another example is the SceneWalk model [8], where a given fixation sequence generates deterministic activation maps for attention and inhibitory tagging. However, in the stochastic SWIFT model [7], there is a fluctuating set of word-based activations that follow the temporal evolution given by the master equation. Thus, in the SWIFT model several runs are necessary to predict the most likely internal state given an experimentally-observed fixation sequence during reading.

In the next section, we will try to classify and compare the types of models that we investigated in this textbook.

8.2 Types of Models in This Textbook

Given the above classification of modeling levels, we might want to take a look back to the model classes discussed in Chaps. 1–7 of this textbook. Here we will follow the classification scheme of mathematical models proposed by Gershenfeld [10], which we adapted to our purposes.

8.2.1 Stochastic Models

Random Variables

In systems with random fluctuations and noise, it is often an important step to specify a random variable that serves as an adequate model of the observed data. As an example, we introduced a temporal Poisson process (Sect. 1.5) to describe the statistical pattern of microsaccade occurrences during fixational eye movements. This model was extended to a time-dependent (or inhomogeneous) Poisson process that could be simulated on a computer. Related to these theoretical concepts, a process model was developed later in Chap. 2.

A random-variable account of data can generate complicated data patterns, e.g., the different versions of spatial point processes discussed in Chap. 6. Such models turned out to be very important as a baseline description for dynamical models of eye guidance during scene viewing.

Finally, random variables can explain complicated processes such as sensorimotor integration (Chap. 4). Using joint and conditional probabilities, the Bayes theorem was applied to motor control, which explained the key effects in oculomotor control during reading, in particular, a mathematical of the launch-site effect was given [19].

Discrete-Time Stochastic Processes

Markov chains were discussed in Chap. 3. Corresponding processes represent the important class of discrete-time stochastic processes with Markov property. We analyzed the simple random-walk model, where analytical solutions could be compared to numerical simulations.

Another, very different model in the class of discrete-time stochastic processes, the time-delayed random-walk, was proposed by Ohira and Milton [20]. While this model produced complicated dependencies, we were still able to derive some analytical solutions for delay $\tau = 0$, which were important for understanding the model's behavior for $\tau > 0$.

Continuous-Time Stochastic Processes

Continuous-time Markov processes are very useful because of their explicit notion of time. This is particularly interesting in neuroscience, since many physiological and neural processes are observed and modeled in continuous time. Therefore, coupling cognitive processes with physiology can be realized in this class of processes in a very straightforward way.

We discussed the master equation for the temporal evolution of probability densities over the discrete state space (Sect. A.2). The master equation is not only a key theoretical concept, but also the theoretical starting point for exact numerical simulations in the *minimal process method* or *Gillespie* algorithm [11].

8.2.2 Deterministic Models

Ordinary Differential Equations and Difference Equations

Ordinary differential equations represent probably the most important class of deterministic models, which we introduced in Chap. 7. Ordinary differential equations may be looked upon as the prototype class of dynamical systems.

In our course, nonlinear oscillations turned out to be an important model class for the description of rhythmic finger movements. While there is clearly fluctuations and noise in these movements, the main effects of rhythmic movements can be explained without a stochastic model.

In the Haken-Kelso-Bunz model [12], two coupled self-sustained oscillators were developed as a model for a qualitative transition in rhythmic finger movements. The coupling term in this model was derived with the help of a potential model that described the stability properties of two dynamical movement states, the symmetrical and anti-symmetrical modes. An important remark is that ordinary differential equation can often be represented as difference equations (which are discrete in time) using a stroboscopic or Poincaré map [13, 14].

Partial Differential Equations

While we did not discuss this class of models, we have to mention partial differential equations for modeling spatially-extended dynamical processes. There are few examples of partial differential equation models in cognitive science, one of the best examples in cognitive modeling is given by the dynamic field theory of movement preparation [2, 9], which was motivated by earlier work on neural activation dynamics [1].

Cellular Automata

Cellular automata are classified as unusual mathematical models [10]. The key concept in a cellular automaton is the storage of the actual state of the model on a lattice combined with a local updating rule. We saw in Sect. 2.5 that such a local updating rule can explain complicated statistical dependencies, in our example the interaction between slow drift motion and microsaccades [6].

While these classes of models do not represent a complete picture of dynamical cognitive modeling (our selection might seem quite arbitrary to some readers), they might provide a good basis for starting with dynamical modeling of eye movements.

8.3 Uses and Abuses of Mathematical Models

In current cognitive science, mathematics plays an important role at all levels [3, 17]: From experimental design and statistical modeling to separate observed effects from noise to knowledge discovery based on machine learning techniques and, finally, to process models derived from first principles. Mathematics requires the scientist to formulate assumptions precisely. Implementing process-based models on a computer is only possible, if the description is unambiguous and clear. Thus, the benefit of mathematics is obvious on all levels of doing research.

The increasing use of complex computer-based models, however, might be problematic, since it unlinks analytical clarity from numerical simulations. Moreover, the availability of increasing computer speed and ease of use of complex simulation environments provide a complicated tool to the non-expert. As a consequence, there is an inherent danger that conclusions from computer-based models are drawn without a clear understanding of the underlying principles and their numerical implementations. According to May [18], one reason for this development is due to the fact that mathematical applications in the life sciences do not have the long-standing tradition of the physical sciences and engineering. As a consequence, we find very useful applications of mathematics, but also abuses [18]:

> Perhaps most common among abuses, and not always easy to recognize, are situations where mathematical models are constructed with an excruciating abundance of detail in some aspects, whilst other important facets of the problem are misty or a vital parameter is uncertain to within, at best, an order of magnitude. It makes no sense to convey a beguiling sense of "reality" with irrelevant detail, when other equally important factors can only be guessed at.

Therefore, training in applied mathematics is essential in our field of cognitive science, where computer-based models represent a substantial part of theories. In many cases, the combination of a simplified model with analytical (approximative) results will lead to a better understanding than the sole application of a complex numerical simulation model.

Finally, most experimental as well as theoretical scientists are trained in a specific area of applied mathematics. Therefore, there will be a natural tendency to investigate a problem using the learned methods within a specific area of mathematics. Such a tendency, however, counteracts the identification of the optimal model type for a given problem. From this perspective, collaboration between scientists from several fields of experimental and mathematical areas is most desirable.

References

1. Amari, S.-I. (1977). Dynamics of pattern formation in lateral-inhibition type neural fields. *Biological Cybernetics, 27*(2), 77–87. https://doi.org/10.1007/BF00337259
2. Beer, R. D. (2000). Dynamical approaches to cognitive science. *Trends in Cognitive Sciences, 4*(3), 91–99. https://doi.org/10.1016/S1364-6613(99)01440-0
3. Busemeyer, J. R., & Diederich, A. (2010). *Cognitive modeling*. Sage. https://uk.sagepub.com/en-gb/eur/cognitive-modeling/Book226030
4. Einstein, A. (1905). Über die von der molekularkinetischen Theorie der Wärme geforderte Bewegung von in ruhenden Flüssigkeiten suspendierten Teilchen. *Annalen der Physik, 322*(8), 549–560. http://onlinelibrary.wiley.com/doi/10.1002/andp.19053220806/epdf
5. Engbert, R., & Krügel, A. (2010). Readers use Bayesian estimation for eye movement control. *Psychological Science, 21*(3), 366–371. https://doi.org/10.1177/0956797610362060
6. Engbert, R., Mergenthaler, K., Sinn, P., & Pikovsky, A. (2011). An integrated model of fixational eye movements and microsaccades. *Proceedings of the National Academy of Sciences, 108*(39), E765–E770. https://doi.org/10.1073/pnas.1102730108
7. Engbert, R., Nuthmann, A., Richter, E. M., & Kliegl, R. (2005). SWIFT: A dynamical model of saccade generation during reading. *Psychological Review, 112*(4), 777–813. https://doi.org/10.1037/0033-295X.112.4.777
8. Engbert, R., Trukenbrod, H. A., Barthelmé, S., & Wichmann, F. A. (2015). Spatial statistics and attentional dynamics in scene viewing. *Journal of Vision, 15*(1), 14, 1–17. https://doi.org/10.1167/15.1.14
9. Erlhagen, W., & Schöner, G. (2002). Dynamic field theory of movement preparation. *Psychological Review, 109*(3), 545–572. https://doi.org/10.1037/0033-295X.109.3.545
10. Gershenfeld, N. A. (1999). *The nature of mathematical modeling*. Cambridge University Press. http://fab.cba.mit.edu/classes/864.17/
11. Gillespie, D. T. (1976). A general method for numerically simulating the stochastic time evolution of coupled chemical reactions. *Journal of Computational Physics, 22*(4), 403–434. https://doi.org/10.1016/0021-9991(76)90041-3
12. Haken, H., Kelso, J. S., & Bunz, H. (1985). A theoretical model of phase transitions in human hand movements. *Biological Cybernetics, 51*(5), 347–356. https://doi.org/10.1007/BF00336922
13. Jackson, E. A. (1992). *Perspectives of nonlinear dynamics* (Vol. 1+2). Cambridge: Cambridge University Press.
14. Kaplan, D., & Glass, L. (2012). *Understanding nonlinear dynamics*. New York: Springer. https://link.springer.com/book/10.1007/978-1-4612-0823-5
15. Klafter, J., & Sokolov, I. M. (2011). *First steps in random walks: from tools to applications*. Oxford: Oxford University Press. https://global.oup.com/academic/product/first-steps-in-random-walks-9780199234868
16. Kliegl, R., & Engbert, R. (2005). Fixation durations before word skipping in reading. *Psychonomic Bulletin & Review, 12*(1), 132–138. https://doi.org/10.3758/BF03196358
17. Lewandowsky, S., & Farrell, S. (2010). *Computational modeling in cognition: Principles and practice*. SAGE publications. https://sk.sagepub.com/books/computational-modeling-in-cognition

18. May, R. M. (2004). Uses and abuses of mathematics in biology. *Science, 303*(5659), 790–793. https://doi.org/10.1126/science.1094442
19. McConkie, G. W., Kerr, P. W., Reddix, M. D., & Zola, D. (1988). Eye movement control during reading: I. the location of initial eye fixations on words. *Vision Research, 28*(10), 1107–1118. https://doi.org/10.1016/0042-6989(88)90137-X
20. Ohira, T., & Milton, J. G. (1995). Delayed random walks. *Physical Review E, 52*(3), 3277. https://doi.org/10.1103/PhysRevE.52.3277
21. Ratcliff, R. (1978). A theory of memory retrieval. *Psychological Review, 85*(2), 59–108. https://doi.org/10.1037/0033-295X.85.2.59
22. Reichle, E. D., Pollatsek, A., Fisher, D. L., & Rayner, K. (1998). Toward a model of eye movement control in reading. *Psychological Review, 105*(1), 125–157. http://dx.doi.org/10.1037/0033-295X.105.1.125

Appendix A
Basic Concepts on Probability and Stochastic Processes

A.1 Probability and Random Variables

A.1.1 Outcomes, Events, and Sample Space

Probability theory is the basic mathematical framework to investigate random events.[1] The *outcome* of an observation of a random event is denoted by ω. The *sample space* is the set Ω whose elements ω represent all possible outcomes of the observation. When tossing a die, the possible outcomes are $\omega = 1, 2, 3, 4, 5, 6$ and the sample space is $\Omega = \{1, 2, 3, 4, 5, 6\}$. If you throw a dart at a large wall, then the outcome is a position on the plane $\omega = (x, y)$ with coordinates x and y and $\Omega = \mathbb{R}^2$. The distribution of the dart's landing position might be approximated by a 2D normal (Gaussian) distribution around target point $(0, 0)$.

Any subset A of the sample space Ω represents a possible event. This is expressed by the statement that an outcome ω *realizes* the event A if $\omega \in A$; otherwise the outcome realizes the complement $\bar{A}$. For events A and B we can consider the union $A \cup B$ of events A and B and their intersection $A \cap B$.

A.1.2 Probability and Empirical Frequency

The probability is a number that is assigned to each event in probability theory. Axiomatically, the collection $\mathcal{F}$ of events is not always identical to the collection of all subsets of Ω for technical reasons. If the sample space Ω is finite or countable, then any subset of Ω is considered to be an event. The probability of event A

[1] We follow the discussion of Brémaud [1] and Taylor and Karlin [7].

R. Engbert, *Dynamical Models In Neurocognitive Psychology*,
https://doi.org/10.1007/978-3-030-67299-7

is written $P(A)$. For the certain event Ω, the probability is $P(\Omega) = 1$, for the impossible event $\emptyset$, the probability is $P(\emptyset) = 0$.

The probability quantifies the likeliness of occurrence for an event. A probability measure must satisfy the axioms of probability. A probability measure defined on $(\Omega, \mathcal{F})$ is a mapping $P : \mathcal{F} \to \mathbb{R}$ such that

1. $0 \leq P(A) \leq 1$.
2. $P(\Omega) = 1$.
3. $P(\sum_{k=1}^{\infty} A_k) = \sum_{k=1}^{\infty} P(A_k)$.

Property 3 is termed sigma additivity. The triple $(\Omega, \mathcal{F}, P)$ is called a probability space.

For application to experimental data, the probability $P(A)$ is interpreted as the empirical frequency of occurrence of the event A. If a sufficiently large number n of observations is made and n_A is the number of realizations of event A, then the empirical relative frequency $F(A)$ approximates the probability, i.e.,

$$P(A) \approx F(A) = \frac{n_A}{n} . \tag{A.1}$$

In the case of tossing a die, the probability $P(A)$ of event $A \subset \{1, 2, 3, 4, 5, 6\}$ is given as

$$P(A) = \frac{|A|}{6} , \tag{A.2}$$

where $|A|$ is the number of elements in the event A. In the example of throwing a dart, it is possible to assume a 2D normal distribution of the dart's landing position. In this case, the probability $P(A)$ can be calculated as

$$P(A) = \frac{1}{2\pi\sigma^2} \int\int_A e^{-\frac{x^2+y^2}{2\sigma^2}} \mathrm{d}x\mathrm{d}y . \tag{A.3}$$

A.1.3 Independence and Conditional Probability

An important question is how different events are related. Two events A and B are said to be *(statistically) independent*, if

$$P(A \cap B) = P(A)\, P(B) . \tag{A.4}$$

Dependence between A and B occurs if $P(A \cap B) \neq P(A)P(B)$. The *conditional probability* of A given B is defined as

$$P(A|B) = \frac{P(A \cap B)}{P(B)} , \tag{A.5}$$

which is defined when $P(B) > 0$. The notion of conditional probability is tightly related to statistical independence. If we have information that B is realized, then $P(A|B)$ quantifies our expectation about A. Obviously, the observation of B is not informative, when A and B are independent, i.e., $P(A|B) = P(A)$ for independent events.

A.1.4 Random Variables

A *random variable* is a function $X : \Omega \to \mathbb{R}$ such that for all $x \in \mathbb{R}$, the event $\{X \leq x\} = \{\omega; X(\omega) \leq x\}$ can be assigned a probability. For discrete random variables we can write the event as $\{X = x\}$ for an arbitrary $x \in \mathbb{R}$. Random variables are often called *random numbers*. Note that it is usually the convention to denote the random variables by capital letters such as X, Y, Z and real numbers by lowercase letters as x, y, z.

If we vary the real number x, then the probability determines a function

$$F(x) = P(X \leq x) \qquad \text{for} \qquad -\infty < x < \infty\,, \tag{A.6}$$

which is called the *(probability) distribution function* of the random variable X. The distribution function for a discrete random variable is a step function,

$$F(x) = \sum_{x_i \leq x} p(x_i)\,, \tag{A.7}$$

however, for a continuous random variable X, the distribution function is continuous and $P(X = x) = 0$ for every value of x. If there is a nonnegative function $f(x) = f_X(x)$ for $-\infty < x < \infty$ with the property

$$P(a < X \leq b) = \int_a^b f(y)\mathrm{d}x \qquad \text{for} \qquad -\infty < a < b < \infty\,, \tag{A.8}$$

then the function $f(x)$ is called the probability density function for the random variable X. The relation to the distribution function is

$$F(x) = \int_{-\infty}^{x} f(\xi)\mathrm{d}\xi\,, \qquad -\infty < x < \infty\,. \tag{A.9}$$

In the case that the distribution function is differentiable with respect to x, the probability density function $f_X(x)$ can be obtained by differentiation, i.e.,

$$f(x) = \frac{\mathrm{d}}{\mathrm{d}x}F(x) = F'(x)\,. \tag{A.10}$$

Therefore, the informal statement

$$P(x < X \leq x + \mathrm{d}x) = F(x + \mathrm{d}x) - F(x) = \mathrm{d}F(x) = f(x)\mathrm{d}x \tag{A.11}$$

is often used.

A.1.5 Expected Values and Moments

For a continuous random variable X with probability density function $f_X(x)$ we can formulate the expectation of a function $g(X)$ by

$$E[g(X)] = \int g(x) f_X(x)\mathrm{d}x \ . \tag{A.12}$$

For a discrete random variable X with values $x_1, x_2, x_3, \ldots$ the expectation can be written in the form

$$E[g(X)] = \sum_{i=1}^{\infty} g(x_i) p_X(x_i). \tag{A.13}$$

Moments are important parameters to characterize a random variable. For a continuous random variable X with probability density $f_X(x)$, the mth moment is defined as

$$E[X^m] = \int_{-\infty}^{\infty} x^m f(x)\mathrm{d}x \ , \tag{A.14}$$

provided that the integral exists. For a discrete random variable X with value x_i (see above) and corresponding probabilities $p_X(x_i)$, the mth moment is given as

$$E[X^m] = \sum_i x_i^m p_X(x_i) \ . \tag{A.15}$$

The *first moment* (i.e., m = 1) is the *mean value* or expected value $E[X] = \mu_X = m_X$. If the mean value exists, then the *variance* is given as the second *central* moment, which is the second moment of the random variable $Y = X - \mu_X$, i.e.,

$$\mathrm{Var}[X] = E[Y^2] = E[(X - \mu_X)^2] = E[X^2] - \mu_X^2 \ . \tag{A.16}$$

A.1.6 Central Limit Theorem

Many random variables that occur in nature are well approximated by a Gaussian distribution. This fact is due to the central limit theorem, which applies to a random variable that Gaussian if it is given as the sum of many statistically independent components with an arbitrary distribution. Consider independent random variables $X_1, X_2, \dots, X_n$ with properties

$$E[X_i] = 0\,, \quad \mathrm{Var}[X_i] = \sigma_i^2\,, \tag{A.17}$$

and probability density $P_i(X_i)$ for $i = 1, 2, \dots, n$. Next, we introduce the sum of the random variables using the definitions

$$S_n = \sum_{i=1}^{n} X_i \quad \text{and} \quad \mathrm{Var}[S_n] = \sigma_n^2 \tag{A.18}$$

and assume that the *Lindeberg Condition* holds, i.e.,

$$\lim_{n\to\infty} \frac{1}{\sigma_n^2} \sum_{i=1}^{n} \int_{|x|>t\sigma_n} x^2 P(x)\mathrm{d}x\,. \tag{A.19}$$

Under these conditions, the distribution of the normalized sums S_n/σ_n tends to the Gaussian with zero mean and unit variance.

A.1.7 Joint and Marginal Distributions

For two random variables X and Y, we can define their *joint probability distribution function* as the function $F_{XY}(x, y)$ of the two real variables x and y,

$$F_{XY}(x, y) = P(X \le x \quad \text{and} \quad Y \le y)\,. \tag{A.20}$$

A joint probability density is a function $f_{XY}(x, y)$ with the property

$$F_{XY}(x, y) = \int_{-\infty}^{x} \int_{-\infty}^{y} f_{XY}(\xi, \eta)\, \mathrm{d}\xi\, \mathrm{d}\eta \tag{A.21}$$

for all x and y. The probability density of a single variable can be obtained from the joint probability density $f_{XY}(x, y)$ via

$$f_X(x) = \int_{-\infty}^{\infty} f(x, y)\, \mathrm{d}y \quad \text{and} \quad f_Y(y) = \int_{-\infty}^{\infty} f(x, y)\, \mathrm{d}x\,, \tag{A.22}$$

where $f_X(x)$ and $f_Y(y)$ are called the *marginal probability distributions* of X and Y, respectively.

A.1.8 *Bayes' Rule*

The conditional probabilities $P(A|B)$ and $P(B|A)$ can be used to symmetrically decompose the joint probability $P(A, B)$ in the following form,

$$P(A \cap B) = P(A|B)\, P(B) = P(B|A)\, P(A) \,. \tag{A.23}$$

In the case that $P(A) > 0$ we obtain Bayes' rule of retrodiction

$$P(B|A) = \frac{P(A|B)\, P(B)}{P(A)} \,, \tag{A.24}$$

which turns out to be an important rule for the computation of complex events.

A.2 Stochastic Processes

In experiments, we will observe a stochastic process in state n_1 at time t_1 with probability $p(n_1, t_1)$. For M observations, state n_1 at time t_1, state n_2 at time $t_2, \ldots,$ state n_M at time t_M, the probability of these M observations is given by the *joint probability*

$$p(n_1, t_1; n_2, t_2; n_3, t_3; \ldots; n_M, t_M) \qquad \text{with} \qquad t_1 > t_2 > t_3 > \ldots > t_M \tag{A.25}$$

which we will write in abbreviated form as $p(1, 2, 3, \ldots, M)$. The definition of the *conditional probability* is given by

$$p(1|2) = \frac{p(1, 2)}{p(2)} \,, \tag{A.26}$$

which is the probability of making observation 1 (state n_1 at time t_1) *given* observation 2, i.e., the process was in state n_2 at time t_2. Using the conditional probability, we can introduce an *uncorrelated process*, which has the property

$$p(1|2, 3, \ldots, M) = p(1) \,, \tag{A.27}$$

i.e., the probability of making observation 1 (state n_1 at time t_1) does not depend on previous observations. Therefore, an uncorrelated process has no memory. A direct consequence of Eq. (A.27), is the factorization of the joint probability,

$$
\begin{aligned}
p(1,2,3,\ldots,M) &= p(1|2,3,\ldots,M)\cdot p(2,3,\ldots,M)\\
&= p(1)\cdot p(2|3,4,\ldots,M)\cdot p(3,4,\ldots,M)\\
&\;\vdots\\
&= p(1)p(2)p(3)\ldots p(M)\,.
\end{aligned}
$$

A.2.1 Continuous-Time Discrete-State Markov Processes

The important class of *Markov processes* has the property that for the conditional probability $p(1|2,3,\ldots,M)$ it holds that

$$p(1|2,3,\ldots,M) = p(1|2)\,, \tag{A.28}$$

which indicates that the probability of making observation 1 (state n_1 at time t_1) depends only on the last of the previous observations, which is, in our case, observation 2 at time $t_2 < t_1$.

In the case of continuous-time discrete-space Markov processes, the temporal evolution of the probability density is described by the *master equation*,

$$\frac{\partial p(n,t|n_0,t_0)}{\partial t} = \sum_{m\neq n}\{W_{nm}p(m,t|n_0,t_0) - W_{mn}p(n,t|n_0,t_0)\}\,, \tag{A.29}$$

with the *initial condition* that the process was in state n_0 at time t_0.[2] The time derivative on the left-hand side of Eq. (A.29) gives the temporal change of the set of probabilities $p(n,t)$ at time t. The right-hand side represents of a balance of probabilities (or *conservation equation*) with *gain* terms $W_{nm}p(m,t|n_0,t_0)$ from all neighboring states m to state n and *loss* terms $W_{mn}p(n,t|n_0,t_0)$ from state n to all neighboring states m. The derivation of the master equation can be found in textbooks on stochastic processes [3, 6, 8]. In the following, we discuss important examples of continuous-time discrete-state Markov processes.

A.2.2 The Random Telegraph Process

The two-state Markov process is often called the *random telegraph process*, a stochastic process with Markov property and two internal states "0", "1", and the corresponding transitions "0"$\mapsto$"1" and "1"$\mapsto$"0". The process can be described by

[2] It should be noted that it is important to use the conditional probability when describing the state of a stochastic process [5] to indicate the dependence on the initial state.

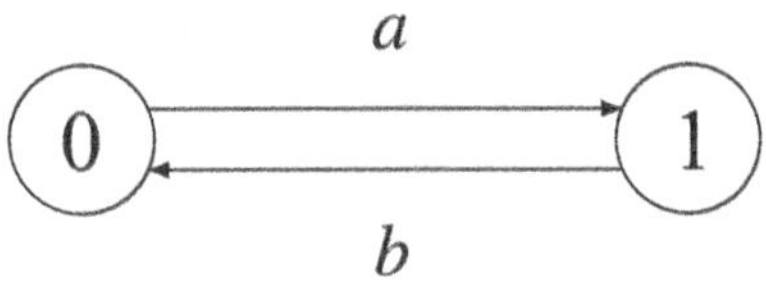

Fig. A.1 The random telegraph process, which is the two-state Markov process. The transition probabilities between the two states are given by parameters $W_{10} = a$ and $W_{01} = b$ of the master equation, Eq. (A.30–A.31)

the diagram in Fig. A.1. The master equation, Eq. (A.29), reduces to a set of two equations for states 0 and 1 with corresponding gain and loss terms, i.e.,

$$\frac{\mathrm{d}p_{0|n_0}}{\mathrm{d}t} = -ap_{0|n_0} + bp_{1|n_0} \tag{A.30}$$

$$\frac{\mathrm{d}p_{1|n_0}}{\mathrm{d}t} = +ap_{0|n_0} - bp_{1|n_0}, \tag{A.31}$$

where the initial condition is that the process starts in state n_0 at time t_0.

Because of the normalization of probability, the sum of the probabilities over the two states 0 and 1 must be normalized to unity, i.e., $p_{0|n_0}(t) + p_{1|n_0}(t) = 1$ for all $t \geq 0$. Transitions between the two states will lead to changes of the probabilities p_0 and p_1, but the normalization has to be conserved over time. Using the master equation, we can directly prove this property by

$$\frac{\mathrm{d}(p_{0|n_0} + p_{1|n_0})}{\mathrm{d}t} = -ap_{0|n_0} + bp_{1|n_0} + ap_{0|n_0} - bp_{1|n_0} = 0\,, \tag{A.32}$$

which shows that the total probability of the process, given by the sum $p_{0|n_0}(t) + p_{1|n_0}(t)$, does not change over time, since the corresponding derivative with respect to time is zero.

The stationary solution of the random telegraph process is the asymptotic probability distribution, which we would obtain after an infinitely long observation time. In this asymptotic equilibrium state of the system, the time derivatives on the left-hand sides of the master equation vanish. Therefore, using the master equation with $\dot{p}_{0|n_0} = \dot{p}_{1|n_0} = 0$, we obtain the relation

$$-a\bar{p}_0 + b\bar{p}_1 = 0\,, \tag{A.33}$$

with the additional requirement of normalization, i.e., $\bar{p}_0 + \bar{p}_1 = 1$. Combining the last two equations, we find that the stationary solution of the random telegraph process is given by

$$\bar{p}_0 = \frac{b}{a+b} \quad \text{and} \quad \bar{p}_1 = \frac{a}{a+b}\,. \tag{A.34}$$

Therefore, the stationary probability distribution of the process is given by the relative transition probabilities.

A.2.3 Simulation: The Minimal Process Method

An efficient method for the numerical simulation of Markov processes can be derived from the master equation. The resulting simulation procedure is known as *minimal process method* or Gillespie algorithm [2, 4]. Let us assume that the process is in state n with certainty, which is "0" or "1" in the case of the random telegraph process. In the case $n = 0$ at time $t = 0$, there is $p_{0|0}(t = 0) = 1$ and $p_{1|0}(t = 0) = 0$. As a consequence, the master equation reduces to

$$\frac{\mathrm{d}p_{0|0}}{\mathrm{d}t} = -ap_{0|0} \,. \tag{A.35}$$

The solution of this differential equation is an exponential function of the form $p_{0|0}(t) = \exp\{-at\}$. The time interval τ to the next transition is given by the derivative of $p_{0|0}(t)$, which leads to the *waiting time* distribution

$$\rho(\tau) = a\mathrm{e}^{-a\tau} \,. \tag{A.36}$$

The waiting time distribution is normalized in the sense that the transition will occur at some point in time $t > 0$, i.e.,

$$\int_0^\infty \rho(\tau)\mathrm{d}(\tau) = \int_0^\infty ae^{-a\tau}\mathrm{d}\tau = \left[-e^{-a\tau}\right]_0^\infty = 0 - (-1) = 1 \,. \tag{A.37}$$

The mean waiting time (expectation) is given by the inverse of the transition probability,

$$E[\tau] = \int_0^\infty \tau\rho(\tau)\mathrm{d}\tau = \frac{1}{a} \,. \tag{A.38}$$

An exponentially distributed variable can be transformed from a computer-based random number generator ξ by

$$\tau = -\frac{1}{a}\log(1-\xi) \qquad \text{with} \qquad 0 \le \xi < 1 \,. \tag{A.39}$$

The minimal process method is highly efficient, since, in a given state, it determines the point in time, where the next transition occurs. The algorithm is exact; a simulated realization is temporally continuous and discretization of time is not necessary (Fig. A.2).

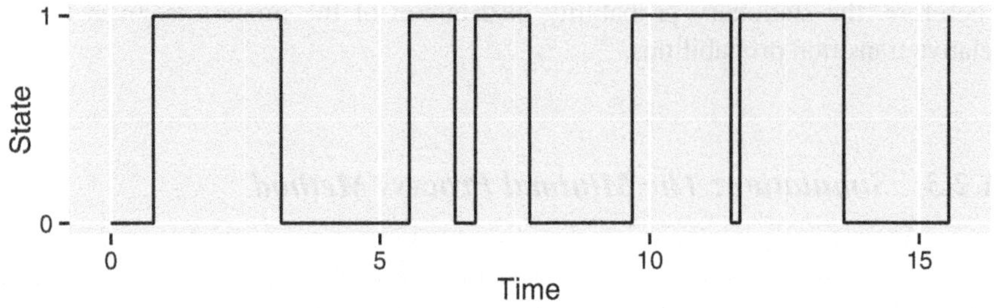

Fig. A.2 Simulation of the random telegraph process using the minimal process method. The irregular distribution of transitions over time indicates that the process and its numerical simulation algorithm is continuous in time

A.2.4 *Relaxation of the RTP to the Stationary State

Based on the master equation for the random telegraph process, Eqs. (A.30–A.31), we can derive an analytical solution for the temporal evolution of the probability density. Using the normalization condition, $p_{0|n_0}(t) + p_{1|n_0}(t) = 1$ for all $t \geq 0$, we can eliminate $p_{1|n_0}(t)$ from Eq. (A.30) and obtain the differential equation

$$\dot{p}_{0|n_0}(t) = -(a+b)\left(p_{0|n_0}(t) - \frac{b}{a+b}\right), \tag{A.40}$$

which has the general solution

$$p_{0|n_0}(t) = \frac{b}{a+b} + C_{n_0}^{(0)} e^{-(a+b)t} \tag{A.41}$$

for $t \geq 0$. The constant $C_{n_0}^{(0)}$ can be determined from the initial conditions, e.g.,

$$p_{0|0}(0) = 1 \quad \text{and} \quad p_{0|1}(0) = 0\,, \tag{A.42}$$

which gives $C_0^{(0)} = a/(a+b)$ and $C_1^{(0)} = -b/(a+b)$. The solution for $p_{1|n_0}(t)$ is derived analogously. In summary, the full solution of the master equation is given by

$$p_{0|0}(t) = \frac{b}{a+b} + \frac{a}{a+b} e^{-(a+b)t}\,, \tag{A.43}$$

$$p_{1|1}(t) = \frac{a}{a+b} + \frac{b}{a+b} e^{-(a+b)t}\,, \tag{A.44}$$

$$p_{0|1}(t) = \frac{b}{a+b}\left(1 - e^{-(a+b)t}\right), \tag{A.45}$$

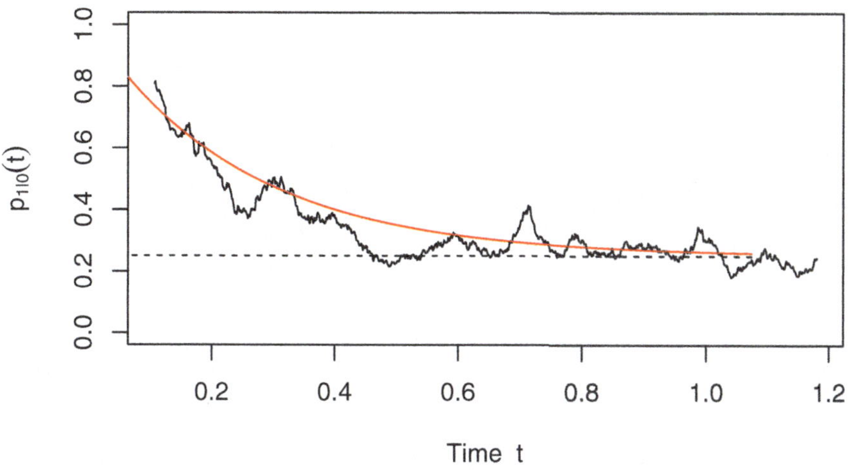

Fig. A.3 Approach of the stationary state. Comparison of the simulated probability $p_{0|0}(t)$ (black line) and the analytical solution (red line), Eq. (A.43), for $a = 1$ and $b = 3$. The simulation is based on $N = 4000$ runs of the random telegraph process

$$p_{1|0}(t) = \frac{a}{a+b}\left(1 - e^{-(a+b)t}\right) . \tag{A.46}$$

To illustrate this behavior via numerical simulation, we choose $n_0 = 0$, simulate $N = 1000$ RTPs, and compute the moving average of the probability $p_{0|0}(t)$ with $p_{0|0}(0) = 1$ and time evolution according to Eq. (A.43). In Fig. A.3, a comparison of the analytical solution for the probability and its numerical simulation is plotted.

A.2.5 Generalization to more than One Adjoined State

For an application of the minimal process method to the continuous-time version of the random-walk model with discrete states (Chap. 3), the minimal process method can be generalized to stochastic processes with several possible transitions to neighboring states. Let us consider the case of a one-dimensional, discrete-state continuous-time random walk. From a given state n at time t, there are two possible transitions, (1) $n \mapsto n+1$ with probability a and (2) $n \mapsto n-1$ with probability b. The corresponding master equation can be written as

$$\frac{\mathrm{d}p_{n|n_0}}{\mathrm{d}t} = ap_{n-1|n_0} + bp_{n+1|n_0} - ap_{n|n_0} - bp_{n|n_0} \tag{A.47}$$

for the temporal evolution of the probability $p_{n|n_0}(t)$ with boundary condition $n(t = 0) = n_0$. Next, we investigate the waiting time distribution, given that the process is

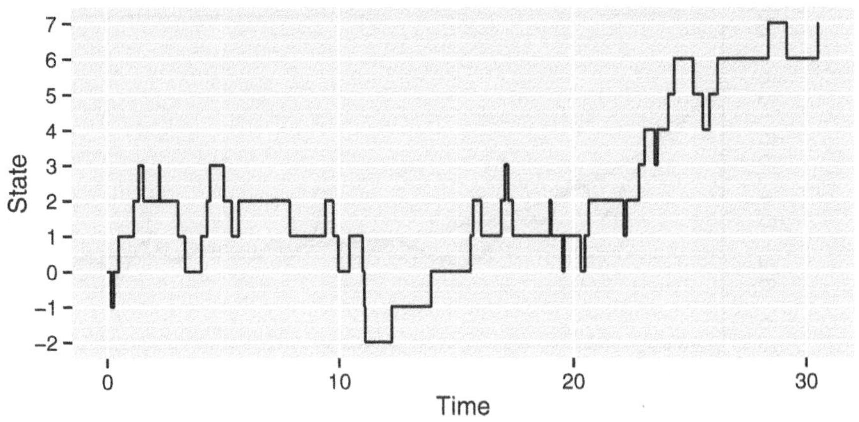

Fig. A.4 The continuous-time random-walk model. Starting from state $n(t = 0) = 0$, the minimal process method simulates two transitions from a given state n at time t, $n \mapsto n+1$ with probability $a = 1$ and $n \mapsto n-1$ with probability $b = 0.8$

in state n at time t with certainty. This assumption leads to the equation $\mathrm{d}p_{n|n}/\mathrm{d}t = -(a+b)p_{n|n}$, since it holds that $p_{n-1|n}(t = 0) = p_{n+1|n}(t = 0) = 0$. In analogy to Eq. (A.36), the waiting time distribution is again an exponential distribution, where the exponent is given by $a + b$ (Fig. A.4).

For general Markov processes, the assumption is that the process is in state n at time t with certainty, i.e., $p_{n|n}(t = 0) = 1$ and $p_{n'|n}(t = 0) = 0$ for all other states n'. This leads to the waiting time distribution in state n, i.e.,

$$\rho(\tau) = W_n \mathrm{e}^{-W_n t} \qquad \text{with} \qquad W_n = \sum_{n \neq m} W_{mn} \, . \tag{A.48}$$

The abbreviation W_n is the sum of the transition probabilities to all adjoined states (total transition probability) that can be reached from state n by a stochastic transition. Thus, the mean waiting time in a given state n can be calculated from the total transition probability W_n by the relation

$$E[\tau] = \frac{1}{W_n} = \left(\sum_{n \neq m} W_{mn} \right)^{-1} , \tag{A.49}$$

which, in the case of the simple random-walk process, gives $\langle \tau \rangle = (a + b)^{-1}$.

After computation of the waiting time, how can we determine the transition in the general case of N adjoining states? To choose a transition between several possible alternatives, transitions are selected randomly according to their relative transition probabilities

$$\pi_j = \frac{W_{jn}}{\sum_{n \neq m} W_{mn}}\,, \tag{A.50}$$

for $j \in \{1, 2, 3, \ldots, N\}$. For the special case of the simple random walk, we obtain

$$\pi_1 = \frac{a}{a+b} \quad \text{and} \quad \pi_2 = \frac{b}{a+b}\,. \tag{A.51}$$

In the case of more than two adjoined states, the stochastic selection from a set of transition probabilities can be implement by the linear selection algorithm (see Sect. 5.4).

References

1. Brémaud, P. (1999). *Markov chains: Gibbs fields, Monte Carlo simulation, and queues* (Vol. 31). Springer. http://www.springer.com/de/book/9780387985091#otherversion=9781441931313
2. Feistel, R. (1977). Betrachtung der Realisierung stochastischer Prozesse aus automatentheoretischer Sicht. *Wiss. Zeitschrift Wilhelm-Pieck Univ. Rostock, 26*, 663.
3. Gardiner, C. (1985). *Handbook of stochastic processes*. Berlin: Springer. http://www.springer.com/de/book/9783540707127
4. Gillespie, D. T. (1976). A general method for numerically simulating the stochastic time evolution of coupled chemical reactions. *Journal of Computational Physics, 22*(4), 403–434. https://doi.org/10.1016/0021-9991(76)90041-3
5. Gillespie, D. T. (1998). Describing the state of a stochastic process. *American Journal of Physics, 66*(6), 533–536. http://dx.doi.org/10.1119/1.18895
6. Schnakenberg, J. (1995). *Algorithmen in der Quantenmechanik und Statistischen Physik*. Ulmen: Zimmermann-Neufang.
7. Taylor, H., & Karlin, S. (1999). *An introduction to stochastic modeling* (third ed.). San Diego: Academic Press. https://www.elsevier.com/books/an-introduction-to-stochastic-modeling/pinsky/978-0-12-381416-6
8. Van Kampen, N. G. (1992). *Stochastic processes in physics and chemistry* (Vol. 1). Amsterdam: Elsevier. https://www.elsevier.com/books/book-series/north-holland-personal-library

Appendix B
Basic Concepts on Ordinary Differential Equations

B.1 Models Based on Differential Equations: Exponential Growth

Differential equations relate a function of a variable to the rate-of-change (derivative) of this variable. In the mathematical description of processes, typically a key variable or observable of the process is under consideration. Since relations between derivatives of variables and the variables of the process are very common, differential equation play a very important role in modeling dynamical phenomena in the natural sciences, economics, and cognitive science.

As a simple case study of a *differential equation*, we study exponential growth. Exponential time dependencies frequently occur in growth processes in biology or economics. For example, the amount of traffic on the internet has been increasing exponentially over many years.[1] If we plot the mean monthly circulation in Megabytes per second (Mbps) over time in the period from May 1998 to October 2001, we observe a dramatic increase (Fig. B.1a). A second plot using a logarithmic ordinate (vertical axis) indicates that the internet traffic growth is an approximately exponential function of time, since we observe a straight line in this semi-logarithmic scaling (Fig. B.1b). Let us denote the traffic at time t with $x(t)$. At some arbitrary starting point $t = 0$ (May 1998 in the case of our time series) we observe an *initial value* $x_0 = x(t = 0)$. For exponentially growth of the traffic $x(t)$, we observe

$$x(t) = x_0 \, e^{at} \,, \tag{B.1}$$

where the constant a is the growth rate parameter. The logarithmic plot in Fig. B.1b leads to a linear relationship, i.e., $\log(x(t)) = at + \log(x_0)$.

[1]This example is adopted from Kaplan and Glass [2].

R. Engbert, *Dynamical Models In Neurocognitive Psychology*,
https://doi.org/10.1007/978-3-030-67299-7

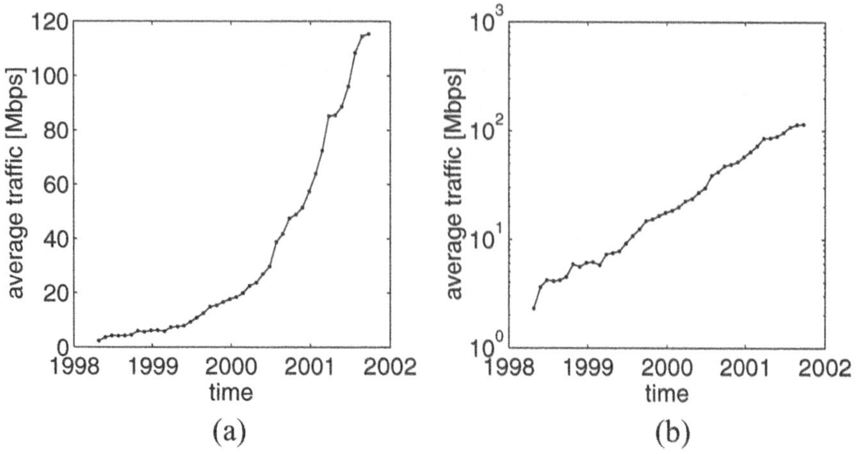

Fig. B.1 Growth of internet traffic. (**a**) Average monthly circulation in [Mbps]) between May 1998 and October 2001. (**b**) Same data as in (**a**), plotted on a logarithmic vertical scale

The rate-of-change of the traffic over time can be calculated directly from Eq. (B.1) by taking the derivative of $x(t)$,

$$\frac{\mathrm{d}x}{\mathrm{d}t} := \lim_{\Delta t \to 0} \frac{x(t + \Delta t) - x(t)}{\Delta t} . \tag{B.2}$$

In our example, we find that the derivative $\mathrm{d}x/\mathrm{d}t$ of the function $x(t)$ is proportional to function $x(t)$ itself, i.e., $\mathrm{d}x(t)/\mathrm{d}t = a\, x_0\, e^{at} = a\, x(t)$, which is a remarkable property that turns out to be characteristic for processes with exponential time-dependence (growth or decay).

The relation between a function and its derivative can be used to formulate an *ordinary differential equation*[2] (ODE) for exponential growth,

$$\dot{x} = \frac{\mathrm{d}x}{\mathrm{d}t} = a\, x , \tag{B.3}$$

where the dot $\dot{x}$ is often used to indicate a derivative with respect to time. We interpret Eq. (B.3) by the statement that the growth rate is proportional to the current amount of traffic $x(t)$ at time t. In biology, a size-dependent growth rate is very common in population dynamics, where the number of offspring is generally proportional to the size of the population. Therefore, we observe exponential growth

[2]The term *ordinary* refers to the property of the equation that one variable and its derivatives are involved, in contrast to *partial differential equations*, where relations between several variables and their derivatives are investigated.

in populations, whose increase is not limited by the resources (e.g., carrying capacity of a habitat).

A solution of Eq. (B.3) is a *function* $x(t)$ *of time* t. In this example, the differential equation can be solved analytically (or algebraically). If we divide both sides by x and multiply with the differential $\mathrm{d}t$, we obtain

$$\frac{\mathrm{d}x}{x} = a\,\mathrm{d}t \ , \tag{B.4}$$

which can be solved formally by integrating both sides,

$$\int_{x_0}^{x(t)} \frac{dx}{x} = \int_0^t a\,\mathrm{d}t \ . \tag{B.5}$$

The solution of the integral on the left-hand side of the equation is the logarithm of x. Therefore, we obtain

$$\log(x(t)) - \log(x_0) = a\,t \ . \tag{B.6}$$

Raising both sides of this equation to the power of the exponential function, we obtain the exact solution of the ODE for exponential growth,

$$x(t) = x_0\,e^{at} \ . \tag{B.7}$$

Note that the same equation produces exponential decay for $a < 0$. For many ODEs, it will turn out to be impossible to solve the integral on the left-hand side of Eq. (B.5) by analytical means. In these cases we can solve the corresponding ODE numerically on a computer to obtain a solution, which is done in Chap. 7.

The ODE for exponential growth, Eq. (B.3), can also be modified, if some external input, in general a function $b(t)$ of time t, is added to the right-hand side of the equation. Here we study the case of exponential decay (negative right-hand side) with additional input, i.e.,

$$\frac{\mathrm{d}y}{\mathrm{d}t} = -a\,y + b(t) \tag{B.8}$$

$$y(t_0) = y_0 \tag{B.9}$$

where, because of the initial value y_0 at time $t = t_0$, the combination of both equations forms an *initial value problem*.

Next, we investigate the analytical solution of the initial value problem for the simple case of constant input, i.e., $b(t) = b = \text{const}$. First, we introduce the substitution

$$u(t) = y(t) - \frac{b}{a} \ , \tag{B.10}$$

which gives the derivative

$$\frac{\mathrm{d}u}{\mathrm{d}t} = \frac{\mathrm{d}y}{\mathrm{d}t} = -ay + b = -a\left(y - \frac{b}{a}\right) = -au\ . \tag{B.11}$$

Therefore, the function $u(t)$ is a solution to the differential equation (B.3) for exponential growth/decay and this solution can be written as

$$u(t) = c \cdot e^{-a(t-t_0)}\ , \tag{B.12}$$

where c is a free parameter. The choice of c is constrained by the need to satisfy the initial value at time t_0,

$$u(t_0) = c\, e^{-a(t_0-t_0)} = c = \left(y_0 - \frac{b}{a}\right)\ , \tag{B.13}$$

which gives, after back-substitution according to Eq. (B.10), the final solution

$$y(t) = \left(y_0 - \frac{b}{a}\right) e^{-a(t-t_0)} + \frac{b}{a}\ . \tag{B.14}$$

Therefore, the resulting solution starts at y_0 at time t_0 and relaxes exponentially to b/a as $t \to \infty$.

B.2 Numerical Solutions of Ordinary Differential Equations

In the SceneWalk model in Chap. 6, we were able to solve the differential equations for the activation maps analytically. In general, this will be impossible for many models. In this case, we must rely on numerical method for the computation of approximative solutions of the corresponding differential equations. First, we motivate the basic concept of numerically solving a differential equation by the Euler method. Second, we continue with an improved method (the Runge-Kutta method), which will be sufficient for our purposes.

B.2.1 Numerical Solutions I: The Euler Method

The starting point for the numerical computation of a solution of an ordinary differential equation (ODE) of the type

$$\frac{\mathrm{d}x}{\mathrm{d}t} = f(x) \tag{B.15}$$

is to approximate the derivative $\mathrm{d}x/\mathrm{d}t$ for *small* time differences Δt by a difference, i.e.,

$$\frac{\mathrm{d}x}{\mathrm{d}t} \approx \frac{x(t+\Delta t) - x(t)}{\Delta t} \,. \tag{B.16}$$

Using this approximation we can rearrange the ODE, Eq. (B.15), to obtain the basic equation for the *Euler method* to compute solutions numerically,

$$x(t+\Delta t) = x(t) + f(x(t))\,\Delta t \,. \tag{B.17}$$

Choosing an *integration time step* Δt, we can iterate the solution of the ODE in steps of Δt forward in time. Let us denote the values of $x(t)$ at times $t = 0, \Delta t, 2\Delta t, 3\Delta t, \ldots$ by $x_0, x_1, x_2, x_3, \ldots$ Using this notation, the algorithm to implement the Euler method can now be written as follows:

1. Start with initial value x_0.
2. Iteration step:

$$x_{n+1} = x_n + f(x_n)\,\Delta t \tag{B.18}$$

3. Return to step 2. until a predefined number of iterations is reached, otherwise stop.

We demonstrate the Euler method for the example of exponential growth,

$$\frac{\mathrm{d}x}{\mathrm{d}t} = a\,x \qquad \text{with solution} \qquad x(t) = x_0\,e^{at} \,, \tag{B.19}$$

with initial value $x_0 = 0.3$, parameter $a = 0.2$, integration step size $\Delta t = 0.5$ over the time interval $[0, 10]$. In Fig. B.2, the red line indicates the numerical solution obtained by the Euler method. The analytical solution (blue line), however, deviates from the numerical solution.

The integration step size Δt is a free parameter of the integration algorithm. A small value of Δt will be necessary to provide a numerical solution, which is close to the exact solution. However, a small value of Δt will be very time-consuming in terms of computer time. How can we control the numerical error? The Euler procedure, Eq. (B.17), can be interpreted as the zeroth-order term of the Taylor expansion of the right-hand side of the ODE. In general, we can increase numerical accuracy by considering additional terms in the Taylor expansion, which is discussed in the next paragraph.

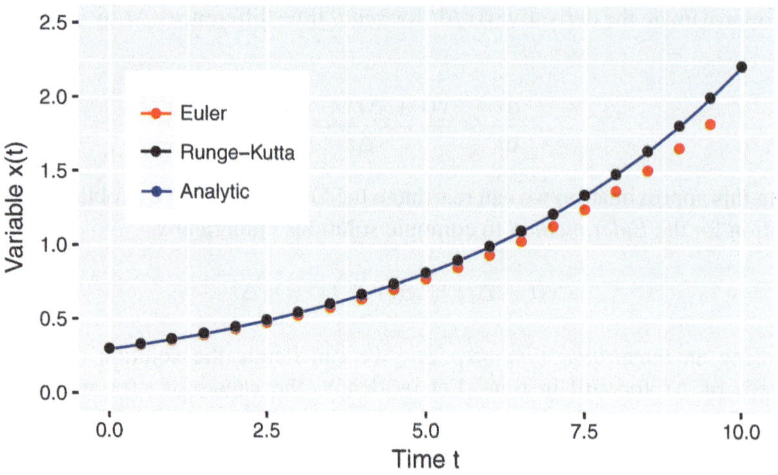

Fig. B.2 Comparison of the Euler and Runge-Kutta methods for numerical integration of the ODE for exponential growth. While the Euler method provide a qualitatively correct solution, there is considerable deviation from the exact (analytical) solution. The fourth-order Runge-Kutta method gives a better approximation of the analytical solution

B.2.2 *Numerical Solutions II: The Runge-Kutta Method*

In the frequently used Runge-Kutta method, a better approximation of the right-hand side of the ODE is used. Let us consider the Taylor expansion of the numerical solution up to second order, i.e.,

$$x(t+h) \approx x(t) + h\frac{\mathrm{d}x}{\mathrm{d}t} + \frac{h^2}{2}\frac{\mathrm{d}^2x}{\mathrm{d}t^2}\,. \tag{B.20}$$

If we replace the derivative by the ODE, $\mathrm{d}x/\mathrm{d}t = f(x)$, we obtain

$$x(t+h) \approx x(t) + h\,f + \frac{h^2}{2}\frac{\mathrm{d}f}{\mathrm{d}t} = x(t) + h\,f + \frac{h^2}{2}\,f\,\frac{\mathrm{d}f}{\mathrm{d}x}\,, \tag{B.21}$$

which can be further simplified by approximating the derivative $\mathrm{d}f/\mathrm{d}x$.

In the *fourth-order* Runge-Kutta method the derivative $\mathrm{d}x/\mathrm{d}t$ is approximated more accurately than in the Euler method; the order refers to the number of steps used in the approximation formula. To compute x_{n+1} from x_n for the fourth-order Runge-Kutta method we have to evaluate the function f at four different values,

$$\begin{aligned}
k_1 &= f(x_n)\,\Delta t\\
k_2 &= f(x_n + 0.5\,k_1)\,\Delta t\\
k_3 &= f(x_n + 0.5\,k_2)\,\Delta t\\
k_4 &= f(x_n + k_3)\,\Delta t\,,
\end{aligned}$$

and the new value x_{n+1} is given by

$$x_{n+1} = x_n + (k_1 + 2k_2 + 2k_3 + k_4)/6 . \quad (B.22)$$

In Fig. B.2, we compare the Euler and Runge-Kutta methods in the case of the ODE for exponential growth. While we saw that there are considerable deviations of the Eulerian solution from the exact result, the Runge-Kutta method gives a good approximation to the exact solution. It should be mentioned that variable step-size routines can improve the accuracy further [3].

B.3 The Harmonic Oscillator

One-dimensional ordinary differential equations are too limited to describe the broad class of oscillatory phenomena, which is the basis for rhythmic finger movements discussed in this chapter. Let us consider a sine-wave oscillation, $x(t) = \sin(\omega t)$. In this case, the *second* derivative of $x(t)$ is proportional to the function $x(t)$ itself, i.e.,

$$\frac{d^2x}{dt^2} = \frac{d}{dt}\left(\frac{dx}{dt}\right) = \frac{d}{dt}(\omega \cos(\omega t)) = -\omega^2 \sin(\omega t) = -\omega^2 x . \quad (B.23)$$

Since there is a second derivative on the left-hand side of the last equation, the equation is called an ODE of second-order. To simplify the analysis, we can rewrite Eq. (B.23) as two ODEs of first-order by substituting a new variable $v = dx/dt$, which describes the velocity of $x(t)$. The resulting equation is the ODE of the *harmonic oscillator*,

$$\frac{dx}{dt} = v \quad (B.24)$$

$$\frac{dv}{dt} = -\omega^2 x . \quad (B.25)$$

For numerical simulations, we apply the Runge-Kutta method *simultaneously* to both equations.

A general solution of the harmonic oscillator equation, Eq. (B.24), is a combination of sine and cosine terms, i.e., $x(t) = A\sin(\omega t) + B\cos(\omega t)$. For an oscillatory process it is not sufficient to specify an initial value $x(t = 0) = x_0$, since different velocities $v(t = 0) = v_0$ will lead to different solutions. Therefore, the state of the oscillator at time t can be represented mathematically by the pair $(x(t), v(t))$ in the *phase space*, which is spanned by two dynamical variables that characterized the process unambiguously. With the knowledge of the state in phase space, we can compute all previous and future states of the system (Fig. B.3).

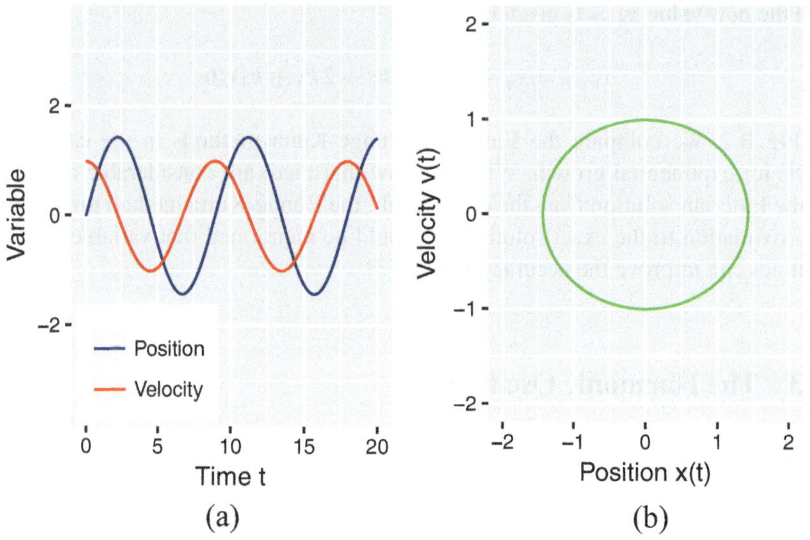

Fig. B.3 Numerical solution of the ODE of the harmonic oscillator. (**a**) A solution $x(t)$ over time t. (**b**) The trajectory of the system in phase space (x, v)

The ODE of the harmonic oscillator is a very important differential equation with many different applications. Here, we show how the harmonic oscillator equation can be derived from a discussion the underlying mechanical forces in the physics of a mass-spring pendulum. If we want to investigate the effects of a force F acting on a small particle of mass m moving at velocity v, we exploit *Newton's second law*, which states that the force is equal to the second time derivative of the position, $\ddot{x}$, multiplied by the mass m,

$$F = m\,a = m\,\ddot{x}\ , \tag{B.26}$$

where $a = \ddot{x}$ is called the acceleration of the particle. In the mass-spring pendulum, the force is generated by the spring. Experiments show that the force F is a linear function of the elongation x, measured as the position x of the pendulum's mass, if the define the equilibrium point of the spring as $x = 0$,

$$F = -k\,x\ . \tag{B.27}$$

This force is opposite in sign with position x, because it acts against the displacement x from the equilibrium position. Combining Eqs. (B.26) and (B.27) and dividing both sides by m gives

$$\ddot{x} + \omega^2 x = 0 \qquad \text{with} \qquad \omega^2 = \frac{k}{m}\ , \tag{B.28}$$

which is, again, the ODE of the harmonic oscillator, Eq. (B.23). The relation of ω and m explains why increasing the mass leads to oscillations with lower frequency.

B.4 Nonlinear Oscillations

If we disturb a trajectory of the harmonic oscillator, the new trajectory will have a different amplitude after the disturbance. However, in many natural *self-sustained oscillators* we found very different behavior: After a perturbation, the system returns to the original amplitude, if we wait sufficiently long. Because almost all initial conditions relax to the oscillation with fixed amplitude and frequency, the corresponding dynamical state is called an *attractor*, in this case a *stable limit cycle*, first discovered by the French mathematician Henri Poincaré (1854–1912) (Fig. B.4).

An important example for a self-sustained oscillatory process is the heart; cardiac oscillations are self-sustained and are, at resting state, stabilized at a fixed frequency and amplitude. An early model for cardiac oscillations, which leads to a stable limit cycle, was proposed by van der Pol and van der Mark in 1928 [1]. The well-known *van-der-Pol equations* are given by

$$\frac{\mathrm{d}x}{\mathrm{d}t} = \frac{1}{\epsilon}\left(y - \frac{x^3}{3} + x\right) \tag{B.29}$$

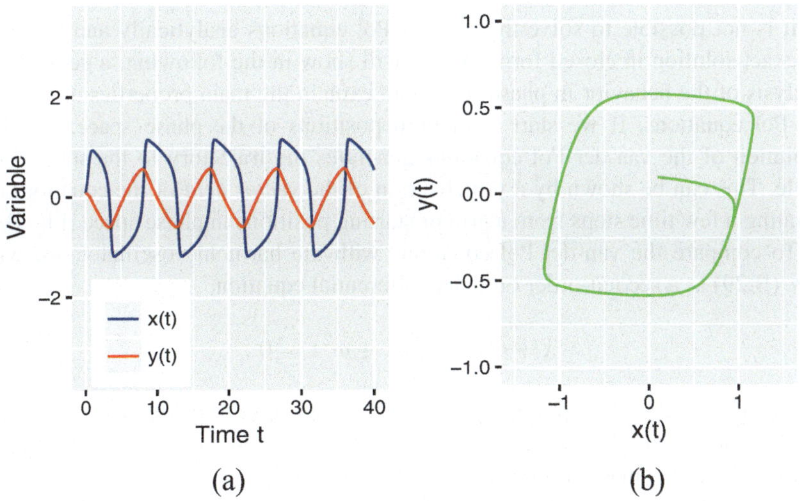

Fig. B.4 Numerical solution of the van-der-Pol oscillator. (**a**) Time series of $x(t)$ and $y(t)$. (**b**) Trajectory $(x(t), y(t))$ in phase space

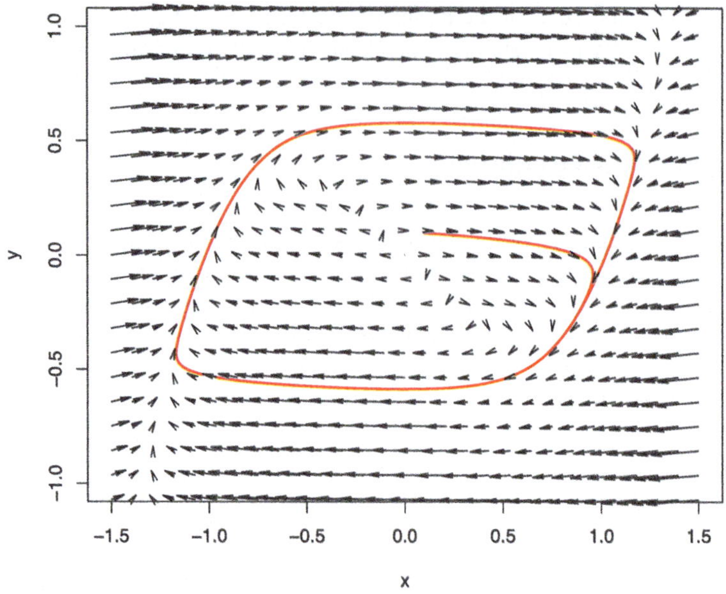

Fig. B.5 Numerical exploration of the vector field of the van-der-Pol equations. The lengths of the streamlines represent the velocity and direction in the corresponding part of the phase space

$$\frac{\mathrm{d}y}{\mathrm{d}t} = -\epsilon\, x \,. \tag{B.30}$$

It is not possible to solve the van-der-Pol equations analytically and to obtain an exact solution in closed form. As we will show in the following, a geometrical analysis of the behavior in phase space can explain the main properties of the van-der-Pol equations. If we start at random positions of the phase space, the time evolution of the van-der-Pol equations generates the trajectory to the stable limit cycle. This can be shown by a visualization of the *vector field* of the equations, by iterating a few time steps from a grid of starting positions in phase space (Fig. B.5).

To compare the van-der-Pol equations with the harmonic oscillator, we write Eqs. (B.29) as a second-order ordinary differential equation,

$$\ddot{x} + \epsilon(x^2 - r_0^2)\dot{x} + \omega^2 x = 0\,, \tag{B.31}$$

where $\omega^2 = \epsilon = r_0^2 = 1$ to obtain Eq. (B.29). The van-der-Pol oscillator represents an extension of the harmonic oscillator, Eq. (B.24), with an additional friction or damping term $\gamma\dot{x}$ that is proportional to the velocity,

$$\ddot{x} + \gamma\dot{x} + \omega^2 x = 0\,. \tag{B.32}$$

Compared to the above simple friction term, however, in the van-der-Pol equations, the friction parameter is a function of the dynamical variable x, i.e.,

$$\gamma(x) = \epsilon(x^2 + r_0^2) . \tag{B.33}$$

If $x^2 > r_0^2$, the contribution of the damping is positive and the oscillator will be damped, until the amplitude r_0 is reached. If $x^2 < r_0^2$, the oscillation will be enhanced. Taking together these two effects, the amplitude r_0 is stabilized in the van-der-Pol equations.

References

1. Guckenheimer, J., & Holmes, P. J. (2013). *Nonlinear oscillations, dynamical systems, and bifurcations of vector fields* (Vol. 42). New York: Springer. http://www.springer.com/de/book/9780387908199
2. Kaplan, D., & Glass, L. (2012). *Understanding nonlinear dynamics*. New York: Springer. https://link.springer.com/book/10.1007/978-1-4612-0823-5
3. Wilson, H. R. (1999). *Spikes, decisions, and actions: Dynamical foundations of neuroscience*. New York: Oxford University Press.

Index

R. Engbert, *Dynamical Models In Neurocognitive Psychology*,
https://doi.org/10.1007/978-3-030-67299-7

The manufacturer's authorised representative in the EU is Springer Nature Customer Service Centre GmbH, Europaplatz 3, 69115 Heidelberg, Germany. If you have any concerns regarding our products, please contact ProductSafety@springernature.com

Printed and bound by CPI Group (UK) Ltd, Croydon, CR0 4YY

30/09/2026

02225665-0010

Dynamical Models In Neurocognitive Psychology